UNDERSTANDING BEHAVIORAL SYNTHESIS

A Practical Guide to High-Level Design

Mentor Graphics® is a registered trademark of Mentor Graphics Corporation.

Synopsys® is a registered trademark of Synopsys, Inc.

Design Compiler™ is a trademark of Synopsys, Inc.

Verilog® is a registered trademark of Cadence Design Systems, Inc.

Monet® is a registered trademark of Mentor Graphics Corporation.

Idea Station® is a registered trademark of Mentor Graphics Corporation.

HiLo® is a registered trademark of GenRad, Inc.

Calma® is a registered trademark of the Calma Company.

All other trademarks mentioned in this book are trademarks of their respective owners.

UNDERSTANDING BEHAVIORAL SYNTHESIS

A Practical Guide to High-Level Design

John P. Elliott
Mentor Graphics, Inc.

KLUWER ACADEMIC PUBLISHERS
Boston / Dordrecht / London

Distributors for North, Central and South America:
Kluwer Academic Publishers
101 Philip Drive
Assinippi Park
Norwell, Massachusetts 02061 USA
Telephone (781) 871-6600
Fax (781) 871-6528
E-Mail <kluwer@wkap.com>

Distributors for all other countries:
Kluwer Academic Publishers Group
Distribution Centre
Post Office Box 322
3300 AH Dordrecht, THE NETHERLANDS
Telephone 31 78 6392 392
Fax 31 78 6546 474
E-Mail <orderdept@wkap.nl>

 Electronic Services <http://www.wkap.nl>

Library of Congress Cataloging-in-Publication Data

A C.I.P. Catalogue record for this book is available
from the Library of Congress.

Consulting Editor: Jonathan Allen, Massachusetts Institute of Technology

Copyright © 1999 by Kluwer Academic Publishers

All rights reserved. No part of this publication may be reproduced, stored in a retrieval system or transmitted in any form or by any means, mechanical, photo-copying, recording, or otherwise, without the prior written permission of the publisher, Kluwer Academic Publishers, 101 Philip Drive, Assinippi Park, Norwell, Massachusetts 02061

Printed on acid-free paper.

Printed in the United States of America

Table of Contents

Preface — xv

Acknowledgements — xvii

1. How Did We Get Here?

Chapter 1 provides a brief history of electronic design automation from IC mask generation to schematic capture to logic synthesis and finally to behavioral synthesis. Extreme market pressures and rapid changes in silicon technology have changed the design process which has in turn led to the development of design tools of increasing complexity.

1.1	Step Back in Time	1
1.2	The Year is 1981...	1
1.3	The Year is 1986...	2
1.4	The Year is 1991...	3
1.5	Today...	4

2. An Introduction to Behavioral Synthesis

Chapter 2 introduces the behavioral design process. The ability to evaluate multiple architectures (which may not be practical in traditional system design) is an important advantage of the behavioral design process.

2.1	**Why Behavioral Synthesis?**	5
	Design Space Exploration	5
	Productivity Gains	7
	Architecture / Goal Flexibility	8
	Limitations of Behavioral Synthesis	8
	Behavioral Design vs. System Design	9
2.2	**The RTL Design Process**	10
	Developing the Algorithm	10
	Exploring Alternate Architectures	11
2.3	**The Behavioral Design Process**	12
	A Simple Algorithm	14
	Data Dependencies	14
	RTL Synthesis Implementation	15

	Resource Allocation	16
	Scheduling	17
	Control Logic / State Machine	18
	Memory	18
	Data Path Elements	19
	Gantt Charts	21
	Chained Operations	22
	Multi-cycle Operations	22
2.4	**Summary**	**23**

3. The Behavioral Synthesis Process

Chapter 3 discusses the steps performed during behavioral synthesis. While this chapter discusses some topics that are not essential to using behavioral synthesis tools, a good understanding of these steps will provide insight into the results produced by behavioral synthesis.

3.1	**The Behavioral Synthesis Process**	**25**
3.2	**Internal Representations**	**26**
3.3	**Resource Allocation**	**27**
3.4	**Scheduling**	**30**
	Scheduling Concepts	30
	ASAP / ALAP Scheduling	30
	Other Scheduling Methods	31
3.5	**Register Allocation**	**35**
	Lifetime Analysis	35
	Register Sharing	36
3.6	**Binding**	**37**
3.7	**Data Path and State Machine Extraction**	**39**
3.8	**Netlisting**	**39**
	Scheduled Design	39
	Final Design	40
3.9	**Summary**	**40**

4. Data Types

Chapter 4 discusses the data types that are supported by behavioral synthesis and how those data types are translated into a gate-level netlist. The job of any synthesis tool is to ultimately translate all data types to those data types that can represent wires.

4.1	Synthesis Considerations	41
4.2	bit / bit_vector Types	42
4.3	boolean Type	43
4.4	std_logic / std_logic_vector / signed / unsigned Types	43
4.5	Integer Type	47
4.6	Enumerated Type	48
4.7	Record Type	49
4.8	Array Type	50
4.9	Types Not Supported for Synthesis	54
4.10	Summary	55

5. Entities, Architectures, and Processes

Chapter 5 discusses the fundamental building blocks of a VHDL description. Designers writing RTL code utilize these building blocks differently than designers writing behavioral code.

5.1	The Entity Declaration	57
	Ports With Abstract Types	58
	Wrapper Models	60
5.2	The Architecture Specification	64
5.3	Processes	65
	Specifying Clock Edges	69
	Timing Diagrams	70
	Scheduling Assumptions	73
5.4	Summary	76

6. Loops

Chapter 6 discusses the role of loops in behavioral descriptions. Unlike RTL synthesis tools, behavioral synthesis tools support FOR loops with both constant and non-constant bounds, WHILE loops, and infinite loops. Behavioral synthesis methodologies encourage the use of loops to allow designers to describe algorithms at a higher level of abstraction.

6.1	Loops in RTL Design	77
6.2	Loop Constructs and State Machines	79
6.3	The EXIT Statement	81
	Specifying a Synchronous Reset	82

	Simulating the Reset Condition	83
	Complex Reset Sequences	85
	Using an Attribute to Specify a Reset Signal	86
	Specifying an Asynchronous Reset	87
6.4	**Types of Loops**	**88**
	Infinite Loops	88
	WHILE Loops	89
	FOR Loops	91
6.5	**The NEXT Statement**	**92**
6.6	**Scheduling Loops**	**94**
	Clock Cycles and Control Steps	95
	Minimum Execution Time	96
6.7	**Loop Unrolling**	**98**
	Complete Unrolling	98
	Partial Unrolling	100
	When to Unroll	101
6.8	**Summary**	**103**

7. I/O Scheduling Modes

Chapter 7 explains I/O scheduling mode terminology, as defined for behavioral synthesis. I/O modes allow the user to determine the relationship between the timing of the behavioral description and the synthesized design.

7.1	**Overview of Scheduling Modes**	**105**
7.2	**Cycle-Fixed Scheduling Mode**	**106**
	Scheduling a Simple Design	106
	Testing Designs Scheduled in Fixed I/O Mode	109
	Scheduling Conditional Branches in Fixed I/O Mode	109
	Scheduling Loops in Fixed I/O Mode	114
	Advantages and Disadvantages of Fixed I/O Mode	117
7.3	**Superstate-Fixed Scheduling Mode**	**118**
	Scheduling a Simple Design	120
	Testing Designs Scheduled in Superstate I/O Mode	121
	Scheduling Conditional Branches in Superstate I/O Mode	124
	Scheduling Loops in Superstate I/O Mode	127
	Advantages and Disadvantages of Superstate I/O Mode	128
7.4	**Free-Floating Scheduling Mode**	**130**
	Scheduling a Simple Design	130

	Testing Designs Scheduled in Free I/O Mode	132
	Advantages and Disadvantages of Free I/O Mode	133
7.5	Summary	134

8. Pipelining

Chapter 8 explains how pipelining can be used to increase the throughput of a design. Pipelining increases the throughput of a design usually at the cost of area, but this is not always necessary if the design is thoroughly analyzed. The use of pipelined components and the pipelining of loops are considered.

8.1	Types of Pipelining	135
8.2	Pipelined Components	135
	Design Benefits	137
	Design Flow Benefits	140
8.3	Loop Pipelining	142
	Scheduling a Pipelined Loop	143
	Latency and Initialization Interval	146
	NEXT and EXIT Statements in a Pipelined Loop	149
	Dependencies in a Pipelined Loop	151
	Simulating a Pipelined Loop	152
	Restrictions on Pipelined Loops	154
8.4	Summary	154

9. Memories

Chapter 9 discusses how arrays can be mapped to memory. Multi-dimensional arrays can be used in behavioral descriptions to manipulate groups of data in a direct, easy-to-understand manner. Arrays can be mapped to a variety of types of memories. This allows the designer can make intelligent decisions about an appropriate memory to include in a design.

9.1	Memories in RTL Design	155
9.2	Mapping Arrays to Memory	155
	Mapping Indices to Addresses	156
	Synchronous Memory	160
	Asynchronous Memory	163
	Memory Ports	166
	Data Dependencies	167
9.4	Summary	172

10. Functions, Procedures, and Packages

Chapter 10 discusses the role of functions, procedures, and packages in behavioral design. When placed in a package, functions and procedures promote reuse, as subprograms in packages can be used in multiple designs. Functions can also be used to simulate the behavior of complex blocks that will be mapped to pre-constructed components during synthesis.

10.1 Subprograms	173
10.2 Functions	174
Mapping to an Operator	174
Preserving a Function	176
10.3 Procedures	177
10.4 Packages	178
10.5 Summary	180

11. Handshaking

Chapter 11 discusses pre- and post-scheduling simulation issues. A design that has been validated at the behavioral level must also be validated after scheduling. The correlation between I/O read and write operations in behavioral designs and scheduled designs is discussed. A simple handshaking protocol is introduced.

11.1 Communication With External Models	181
Scheduling Assumptions	181
Synchronizing Communication With Fixed I/O	182
11.2 Handshaking	184
Full vs. Partial Handshaking	184
Input Handshaking	186
Output Handshaking	190
Scheduling Issues	191
11.3 Interprocess Communication	195
11.4 Summary	196

12. Reusable Test Benches

Chapter 12 discusses how to design a test bench that can be used with a behavioral design, a scheduled RTL design, as well as the resulting optimized gate-level design. The test bench uses the handshaking principals introduced in Chapter 11.

12.1	**Objectives**	**197**
12.2	**I/O Timing**	**198**
12.3	**Interface Type Considerations**	**199**
12.4	**Test Bench Structure**	**202**
	The ENTITY Declaration	202
	Instantiating the Design	203
	Clock Generation	204
	Reset Generation	204
	Input and Output Processes	205
12.5	**Messages**	**210**
	Assertion Statements	210
	Text I/O	210
12.6	**Summary**	**211**

13. Coding For Behavioral Synthesis

Chapter 13 incorporates the discussion of the previous chapters into coding styles for Behavioral Synthesis. While not defining a set of "hard-and-fast" rules, this chapter does develop a set of guidelines would should allow a designer to successfully evaluate a particular piece of behavioral code and understand how it will be processed by a behavioral synthesis tool.

13.1	**Overview**	**213**
13.2	**Entities, Architectures, and Processes**	**214**
	Specifying Clock Edges	215
	Resets	216
13.3	**Data Types**	**218**
	bit / bit_vector Types	219
	boolean Type	219
	std_logic / std_logic_vector / signed / unsigned Types	219
	Integer Type	223
	Enumerated Type	224
	Record Type	224
	Array Type	224
	Types Not Supported for Synthesis	225
13.4	**Coding Style and I/O Scheduling Mode**	**225**
13.5	**Fixed I/O Scheduling Mode**	**225**
	Conditional Statements	227
	Loops	232

13.6 Superstate I/O Scheduling Mode	235
Conditional Statements	238
Loops	239
13.7 Free Scheduling Mode	241
13.8 Summary	243

14. Case Study: JPEG Compression

Chapter 14 discusses the implementation of a JPEG compression algorithm. The JPEG algorithm represents a compression standard developed for grayscale and color continuous-tone still images. At the heart of the JPEG algorithm is a discrete cosine transform (DCT) which is common in video and audio processing algorithms. This chapter discusses the details of the algorithm itself and the synthesis issues that should be considered in order to produce quality results.

14.1 Introduction	245
14.2 The Algorithm	245
Algorithm Overview	246
Processing an Entire Image	247
The Discrete Cosine Transform	247
Quantization	249
Zigzag Sequence	250
Run-Length / Entropy Encoding	251
14.3 The Environment	252
14.4 Compression Results	254
14.5 Behavioral Description	254
Memories	254
Handshaking	255
Matrix Multiplication	255
14.6 Behavioral Synthesis	255
Bounding the Design Space	256
Exploring Other Architectures	260
14.7 Summary	262

15. Case Study: FIR Filter

Chapter 15 discusses the design and implementation of a Finite Impulse Response (FIR) digital filter. Digital filtering is an important aspect of most DSP-oriented designs. This filter can be easily described using a behavioral coding style and the results of behavioral synthesis are easily understood. Various implementations are considered, including two pipelined architectures.

15.1	Introduction	263
15.2	The Algorithm	263
15.3	Behavioral Description	264
	Tap Coefficients	264
	Filtering	265
15.4	The Environment	267
15.5	Behavioral Synthesis	268
	Bounding the Design Space	268
	Pipelining the Design	270
15.6	Summary	272

16. Case Study: Viterbi Decoding

Chapter 16 discusses the implementation of a Viterbi decoding algorithm. The Viterbi decoding algorithm is used to decode convolution codes and is found in almost every system that receives digital data which might contain errors. This chapter discusses the details of the algorithm itself and the synthesis issues that should be considered in order to produce quality results.

16.1	Introduction	273
16.2	The Algorithm	273
	Convolution Codes	274
	Trellis Diagram	275
	Viterbi Decoder	276
16.3	Behavioral Description	278
16.4	The Environment	279
16.5	Decoding Results	280
16.6	Behavioral Synthesis	281
	Bounding the Design Space	281
	Pipelining the Design	283
16.7	Summary	284

Appendix A: JPEG Source Code	285
Appendix B: FIR Filter Source Code	293
Appendix C: Viterbi Source Code	297
Glossary	301
References and Resources	309
CD-ROM	311
CD-ROM Contents	311
CD-ROM License Agreement	312
Index	313

Preface

Today's designs are creating tremendous pressures for digital designers. Not only must they compress more functionality onto a single IC, but this has to be done on shorter schedules to stay ahead in extremely competitive markets. To meet these opposing demands, designers must work at a new, higher level of abstraction to efficiently make the kind of architectural decisions that are critical to the success of today's complex designs. In other words, they must include behavioral design in their flow.

For any design, there are a multitude of possible hardware architectures the design team could use to achieve their goals. Finding a suitable architecture demands a disproportionate amount of time from even the most experienced designer. When using only logic synthesis tools, an RTL description of that architecture must be created in a hardware description language (HDL) with thousands of lines of handcrafted code. The designer must specify all the details in the code, including the logic states and the corresponding operations performed during those states. Unfortunately, defining an architecture and creating RTL code consume so much time, the designer cannot afford to make any architectural changes, forcing the designer to adopt one architecture and hope for the best.

To raise architectural design above these time-consuming details of the RTL level, behavioral design tools were introduced a few years ago. These tools allow the designer to describe system functionality much more efficiently at the behavioral level, using descriptions typically 5-10 times smaller than an RTL description.

A behavioral description defines the algorithm to be performed with few or no implementation details. Designers direct behavioral synthesis tools to generate alternate architectures by modifying constraints (such as clock period, number and type of data path elements, and desired number of clock cycles). Behavioral synthesis tools take a behavioral description and automatically schedule the operations into clock cycles. The tools automatically create the state machine that is implied when the calculation of an algorithm takes multiple clock cycles. The final output is RTL code which describes the data path elements, the finite state machine, and the memory or registers. This RTL code is ready for logic synthesis.

The biggest challenge to adopting behavioral design is changing the mindset of the designer. Instead of describing system functionality in great detail, the designer outlines the design in broader, more abstract terms. The ability to easily and efficiently consider multiple design alternatives over a wide range of cost and performance is an extremely persuasive reason to make this leap to a higher level of abstraction. Designers that learn to think and work at the behavioral level will reap major benefits in the resultant quality of the final design.

But such changes in methodology are difficult to achieve rapidly. Education is essential to making this transition. Many designers will recall the difficulty transitioning from schematic-based design to RTL design. Designers that were new to the technology often felt that they had not been told enough about how synthesis worked and that they were not taught how to effectively write HDL code that would synthesize efficiently.

This book attempts to address these issues for behavioral synthesis. The intended audience is the designer who will be using (or is considering using) behavioral synthesis, the manager (or others) who will be working with those designers, or the engineering student who is studying leading-edge design techniques.

This book is not a theoretical treatise on behavioral synthesis technology nor is it a reference manual for a particular tool. Using this book, a designer will understand what behavioral synthesis tools are doing (and why) and how to effectively describe their designs so that they are appropriately synthesized.

The book was developed based on the capabilities and features of behavioral synthesis tools that are available today. Certain limitations or inefficiencies that are described in this book will likely go away as such tools evolve.

Although behavioral descriptions can be written in both VHDL and Verilog®, this book focuses entirely on using VHDL for creating behavioral descriptions. Books that attempt to simultaneously address design issues for both VHDL and Verilog® rarely do service to either language. It seemed best that each language should be discussed separately. VHDL was selected first.

The book assumes a working knowledge of the VHDL language and of RTL synthesis concepts. Books that can provide background knowledge on these topics are listed in the *References and Resources* section of this book.

With the benefits of today's behavioral synthesis tools, behavioral design will become a mainstay of IC design in the near future. New interactive behavioral synthesis tools that are capable of quickly processing even the largest designs will make this possible. Since these tools are easily integrated into existing flows, they are particularly attractive to designers under pressure to create viable designs on tight schedules. Behavioral design will soon become a widespread practice throughout the design community.

Acknowledgements

I would like to acknowledge the efforts of those individuals who helped in the creation of this book.

First, I would like to thank my wife, Deborah Layne, who carefully read the first drafts of the material and helped identify the conceptual "leaps" that needed further explanation. Also, her help in compiling the index simplified what I considered an arduous task.

Todd Selden provided a detailed review of the book and in particular furnished insightful commentary on technical errors and omissions.

Rick Nixon provided the opportunity to write this book and also reviewed its contents. Michael Philippi was instrumental in bringing the book to publication. I am particularly grateful for their support.

My thanks to Aaik van der Poel for providing the material for Chapter 1.

I would also like to thank the other engineers at Mentor Graphics who took the time to review the contents of this book.

Finally, my thanks to Carl Harris of Kluwer Academic Press for his encouragement and support.

About the Author

John Elliott began his career in EDA in 1988 at Trimeter Technology Incorporated, one of the first synthesis-related start-up companies, which was purchased by Mentor Graphics in 1989. After working at Mentor in various development, marketing, and application positions both here and in Europe, he worked in the Consulting organization at Synopsys, Inc. and returned to Mentor Graphics in 1996. Since returning to Mentor Graphics, he has worked on teams focused on FPGA design flows, internet technology, and behavioral synthesis.

He holds a B. S. in Electrical Engineering and a B. A. in Philosophy from the University of Pittsburgh, where he was a Chancellor Scholar.

He can be reached at john_p_elliott@yahoo.com

Chapter 1

How Did We Get Here?

1.1 Step Back in Time

To understand the development of behavioral synthesis, take a step back in time and relive some of the EDA tool decisions that designers have made in the past. Put yourself in the position of the typical designer and walk through this bit of history.

1.2 The Year is 1981...

The year is 1981 and complex designs are called Large Scale Integrations (LSI). The complexity of these chips is on the order of hundreds of gates. You do not have a PC or a workstation on your desk, but the boss just armed you with the very latest polygon-pushing tool (from Calma®).

You are one happy, productive designer. The day-to-day pressures of time to market, cost of design, design complexity and design performance are quite manageable. And now you can actually verify your design ideas with a batch-oriented simulator called HiLo®.

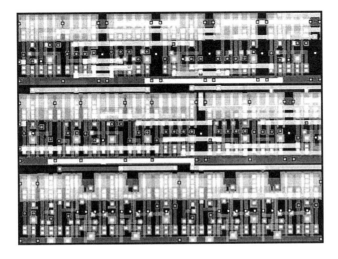

1.3 The Year is 1986...

Well it is 1986 and you have been very successful with these tools for the last 5 years. But now designs have a "V" in front of them (VLSI), and contain thousands of gates.

You are starting to feel pressured. Designs are more complex and market windows are shortening. You barely have time to complete your designs. It has been a while since you had time to sit back and think about how to create more innovative and cost-effective designs. To make matters worse, you are starting to wonder more and more whether you are actually simulating the same design that you are implementing. Now the lack of integration between your entry and verification tools makes you wonder when your entire methodology is going to fall apart.

Time to talk to the boss about some new tools.

Fortunately the boss agrees with your concerns. The next thing you know you have a large box sitting next to your desk -- the boss called it an Apollo® Workstation. On it is a suite of tightly integrated schematic capture and simulation tools. They are not only tightly integrated, but they are also highly interactive. These tools allow you to debug your designs much more efficiently. Mentor Graphics' Idea Station® has come to your rescue.

Since you are now working at a higher level of abstraction you can once again handle the design complexity well within your ever-shortening time to market window. Sure, these designs aren't quite as good as what you did at the mask level, but you can design more quickly, you are sure you are verifying what you are implementing, and redesign costs have gone down dramatically. Time is available again to come up with innovative designs that are better and faster than those of the competition.

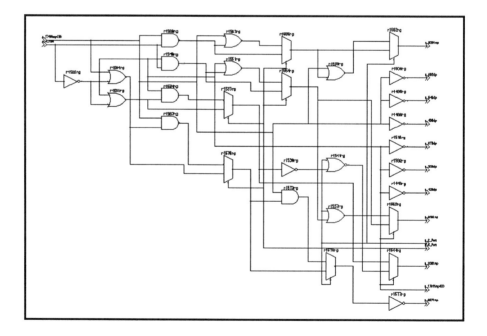

1.4 The Year is 1991...

You had things pretty much under control for the last 5 years. You created some rather complex designs that make you feel pretty proud. But things are changing so fast these days.

Each year ASIC vendors are bringing large numbers of faster and cheaper technologies on-line. This, of course, is also the era of the up and coming FPGA vendors. Vendors are releasing new technologies so quickly, that by the time you have designed your schematics, a better technology is already available. Progress is a good thing, right? Well, you cannot keep changing your schematics indefinitely -- you have to go to production somewhere along the line. But you always feel that you are wasting money on your production runs.

It is becoming harder and harder to keep up with these new technologies. What you need is a technology-independent design methodology. But you don't want to give up the integration with, and the interactivity of, the verification tools to efficiently debug your designs.

You are allowed to attend a Hardware Description Class, and the boss even bought you one of those hot new Synthesis tools – a copy of Design Compiler™ from Synopsys®. No more mind-numbing schematic changes and you finally get to try out some new design ideas that were lingering in the back of your mind for some time now. You describe your design in VHDL or Verilog® and select your target technology. Sure the results aren't quite as good as what you could do at the gate level (if you had time, that is). But if a cheaper technology comes along, all you have to do now is run the design description through again, and you are on your way.

Going to work is fun again!

```
ARCHITECTURE synth OF calc IS                           PROCESS( d, e )
    SIGNAL sub, subreg : signed( 31 DOWNTO 0 );         BEGIN
    SIGNAL mul, mulreg : signed( 31 DOWNTO 0 );             sub <= d - e;
    SIGNAL add, addreg : signed( 31 DOWNTO 0 );         END PROCESS;

    TYPE state_type IS ( s1, s2, s3 );                  PROCESS( a, b, f, subreg, state )
    SIGNAL state, next_state : state_type;                  VARIABLE mul1 : signed( 31 DOWNTO 0 );
BEGIN                                                       VARIABLE mul2 : signed( 31 DOWNTO 0 );
                                                        BEGIN
    PROCESS( state )                                        IF ( state = s1 ) THEN
    BEGIN                                                       mul1 := a;
        CASE state IS                                           mul2 := b;
            WHEN s1 => next_state <= s2;                    ELSE
            WHEN s2 => next_state <= s3;                        mul1 := subreg;
            WHEN s3 => next_state <= s1;                        mul2 := f;
        END CASE;                                           END IF;
    END PROCESS;                                            mul <= mul1 * mul2;
                                                        END PROCESS;
    PROCESS( clk )
    BEGIN                                               PROCESS( a, b, f, yreg, state )
        IF ( clk'EVENT AND clk='1' ) THEN                   VARIABLE add1 : signed( 31 DOWNTO 0 );
            state <= next_state;                        BEGIN
        END IF;                                             IF ( state = s2 ) THEN
    END PROCESS;                                                add1 := c;
                                                            ELSE
    PROCESS( clk )                                              add1 := yreg;
    BEGIN                                                   END IF;
        IF ( clk'EVENT AND clk='1' ) THEN                   add <= add1 + mul_reg;
            subreg <= sub;                              END PROCESS;
            mulreg <= mul;
            addreg <= add;                              y <= addreg;
        END IF;
    END PROCESS;                                        END synth;
```

1.5 Today...

Well, now it is 1999. You have used RTL synthesis for years now, and you have been successful in using different technologies to create an array of innovative products. Since you have solved your Y2K issues, you have time to notice that things have changed again.

RTL synthesis has allowed designs to increase in complexity. (Will this ever end?) Not only have the size of the designs increased, but the complexity of the algorithms that you are implementing has also increased.

Other things have changed. You are no longer responsible for just your own design – you are managing the results of a ten-person design team. You are feeling those old familiar pressures from the days of schematic entry.

You want to understand the causes for the pressures a little better, so you call a meeting to see what the rest of your team is experiencing. You open the meeting with your current project specifications. You put your current RTL flow on the table and ask everyone for input as you go through it. The results confirm your fears: the work is too labor intensive.

There is very little time left to look at innovative architectures. Even if the chosen architecture falls within overall design specifications, your engineers tell you that there is no time left to look for better alternatives. This is no way to guard a leading market position. Just like when you were doing schematic capture and the technologies were changing quickly, you are once again wasting money because you cannot explore better, more cost-effective architectures.

The team concludes that they need to be able to explore multiple architectures *before* committing to the RTL code and the implementation. They need a tool or methodology that allows them to capture design ideas in an *architecture-independent* fashion just like RTL captured ideas in a *technology-independent* fashion. They also need to better understand the implications of design partitioning and to be able to rapidly evaluate constraints between partitions.

Somebody mentions behavioral-level design...

Chapter 2

An Introduction to Behavioral Synthesis

2.1 Why Behavioral Synthesis?

Imagine the following excerpt from a brochure about the latest EDA technology...

> ...Today's ASIC and system-on-chip designers face an unprecedented productivity challenge. Designs are becoming unmanageably complex, time to market pressures are increasing, and ASIC designers are in short supply. Current methodologies are breaking under the strain. Designers must work at a new, higher level of abstraction to make the decisions that are critical to the success of today's complex designs. They must make these decisions quickly and confidently to deliver high performance designs on schedule...

What year is it: 1990 or 1999? Increased design complexity and compressed design cycles have been the reality for hardware designers for some time. In the early 1990's, the introduction of RTL synthesis tools allowed designers to move away from the details of technology-specific schematics and instead describe functionality at a more abstract, technology-independent level using new languages like VHDL and Verilog®.

> In the same way that RTL design allows hardware designers to work at a higher level of abstraction than schematic design, so does behavioral design raise the abstraction level even higher.

2.1.1 Design Space Exploration

For any design, there is a multitude of possible hardware architectures the design team could use to achieve their goals. Finding a suitable architecture demands a disproportionate amount of time from even the most experienced designer. When using only logic synthesis tools, an RTL description of that architecture must be created in a hardware description language (HDL) typically containing thousands of lines of handcrafted code. The designer must specify all the details in the code, including the logic states and the corresponding operations performed during those states.

> *Unfortunately, defining an architecture and creating the corresponding RTL code consume so much time the designer cannot afford to make any architectural changes. The designer is forced to adopt one architecture and hope for the best.*

A typical design will contain a set number of data path elements (such as multipliers and adders). A finite state machine controls the use of these resources, and intermediate values are stored in a memory or in registers. In an RTL description, all of these elements are explicitly defined.

When an RTL description is processed using an RTL synthesis tool, design space exploration is limited. The designer can make some small trade-off between area and delay (as shown in Figure 2-1), but the overall architecture of the design must be explicitly specified in the source code. To explore a different area of the area-delay curve, the RTL description must be re-written. The process of re-writing the RTL code is slow and prone to error. It may be weeks before a re-write has been completed with no guarantee of superior results.

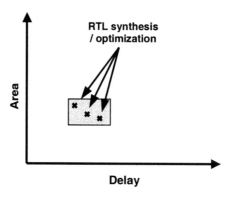

Figure 2-1: Design space exploration is limited with existing RTL design methodologies

Unlike an RTL-based design methodology, a behavioral design methodology shifts the focus from the implementation details to the overall behavior of the design. Design teams are able to think about the design at the highest, algorithmic level. A behavioral description defines the algorithm to be performed with few or no implementation details. For this reason, behavioral descriptions are typically 5-10 times smaller than an equivalent RTL description.

Designers direct behavioral synthesis tools to generate alternate architectures by modifying constraints (such as clock period, number and type of data path elements, and desired number of clock cycles). Behavioral synthesis tools take a behavioral description

An Introduction to Behavioral Synthesis

and automatically schedule the operations into clock cycles. The final output is RTL code which describes the data path elements, the finite state machine, and the memory or registers. This RTL code is ready for logic synthesis. These RTL descriptions can be optimized in a manner similar to a hand-written RTL description, as shown in Figure 2-2, but the architectures that are the starting points for these optimizations are vastly different.

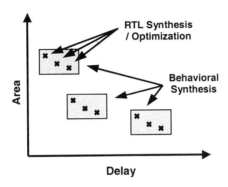

Figure 2-2: Design space exploration with a behavioral design methodology

A behavioral design methodology enables designers to quickly explore many different architectures to ensure that the design constraints can be met before going to RTL synthesis.

> *Using a behavioral design methodology, designers can make well-founded architectural decisions early in the design process.*

Designing at the behavioral level better ties the algorithm specification to the implementation. In many companies that utilize RTL design practices, writing a behavioral model is considered a "luxury". And even when such a model is written, there is no guarantee that it will have a strong correlation to the RTL model.

2.1.2 Productivity Gains

There are benefits associated with describing designs at a higher level of abstraction.

High-level specifications:

- are smaller and less complex than equivalent RTL descriptions
- are easier to write since they contain few implementation details
- are easier to understand and debug since they are more closely related to the algorithm being developed
- are faster to simulate (typically) than an equivalent RTL description, promoting early system verification

2.1.3 Architecture / Goal Flexibility

Behavioral descriptions are *architecture-independent*. Unlike RTL descriptions, behavioral descriptions contain little or no information about the architecture of the final implementation. This means that a single behavioral description can be developed to satisfy a potentially wide variety of goals. A behavioral synthesis tool can create a number of different architectures according to the needs of the design team. Goals that might affect the implementation are:

- Unknown design space

    When implementing new functionality, the architecture that best suites a particular goal may be unknown. When implementing such a design using an RTL methodology, it may not be possible to determine that an architecture is inadequate until late in the design cycle.

- Time-to-market

    Architectural exploration is time-consuming when an RTL design methodology is used. Behavioral synthesis allows a designer to quickly describe and validate the desired functionality of the design and use an architectural exploration tool to select the appropriate architecture.

- Reuse

    Because behavioral descriptions are architecture-independent, a description that is designed with one set of design goals can be easily reused, even when the goals are entirely different.

2.1.4 Limitations of Behavioral Synthesis

Behavioral synthesis should not be used to address every design problem. While a general-purpose behavioral synthesis tool can address many design problems in a satisfactory manner, certain designs may be better implemented with different or more specialized tools.

Behavioral synthesis should not be used or is less effective when:

- The design is not synchronous.

    Behavioral synthesis tools, like RTL synthesis tools, produce synchronous designs. Asynchronous design is not addressed by this technology. But like RTL descriptions, simple asynchronous logic (such as gated clocks) can also be specified in behavioral descriptions.

- The required architecture (structure) for the design is well understood.

    A behavioral description defines an algorithm to be constructed with few or no implementation details. Designers direct behavioral synthesis tools to generate alternate architectures by modifying constraints. If the desired structure of a design is already known, it may be more efficient to describe the design at the RTL level, rather than attempt to develop a more abstract behavioral description.

An Introduction to Behavioral Synthesis

- The design is expected to be near the limits of the area and / or performance capabilities of the target technology and the architecture of the extreme cases is known.

  The flexibility obtained with behavioral synthesis tool has some "cost" in terms of the area and performance characteristics of a design. While a design that is obtained using behavioral synthesis may not be significantly larger than an equivalent hand-coded RTL design, the added area or delay may be unacceptable for extremely high-performance designs.

2.1.5 Behavioral Design vs. System Design

Behavioral design is different from system design, but the line that separates hardware design from system design is becoming blurred. "System-on-chip" (SOC) designs, which may contain memory, programmable cores, and other intellectual property (such as an MPEG encoder) are becoming increasingly common. The overall architecture of such a chip is shown in Figure 2-3.

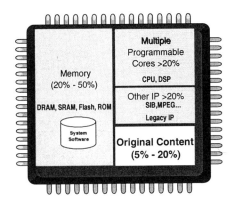

Figure 2-3: Block diagram of a typical System-On-Chip (SOC)

"System design" usually refers to analysis that is beyond the scope of behavioral design, such as hardware-software tradeoffs. Since behavioral synthesis provides the capability to quickly generate multiple architectures early in the design cycle, the detailed analysis of such issues (although not automated) is much more realistic. Behavioral synthesis tools in the future may help make system-level decisions in a more automated manner, but today, behavioral synthesis is limited to the hardware implementation aspects of system level design.

> *Behavioral design is applied to the application specific portion of a given design effort -- the same portion of the chip for which RTL design tools are used.*

The original content is described in terms of a behavioral specification, usually written in either VHDL or Verilog®. Some behavioral synthesis tools support a form of "C", modified to support parallelism and bit-level operations.

To introduce the initial concepts associated with behavioral synthesis, the remainder of this chapter examines a very simple design, discussing the steps that would be performed *manually* by a designer using traditional RTL synthesis tools. This chapter will show that even for a very simple design, there are several different possible architectures, each of which may address different design goals. Performing the steps manually on a simple design will demonstrate how time consuming and difficult the evaluation of multiple architectures for a "complex" design can be. Terms associated with behavioral synthesis are also introduced throughout this discussion.

2.2 The RTL Design Process

Figure 2-4 shows a very high-level depiction of a design process that utilizes Register Transfer Level (RTL) synthesis.

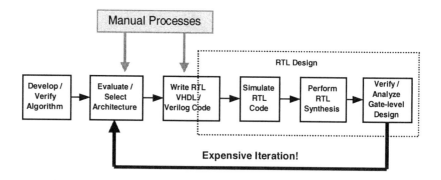

Figure 2-4: The traditional RTL design process

The last three steps in this flow are relatively well understood. RTL synthesis and gate-level verification have been employed in the electronics industry for a number of years and are not discussed in this book.

2.2.1 Developing the Algorithm

The first step in the process, developing the algorithm to be implemented, varies widely from one industry segment to the next. In some industries this is a very formal process that involves a design team that is entirely separate from the implementation team. This is often the case when a company is investigating new approaches to a design problem, and standards are in a state of flux. In other industries, where standards have solidified, a design team might be implementing a well defined or public-domain algorithm in which case it is usually unnecessary to have a separate team to investigate algorithmic alternatives. The processes that are used to develop algorithms, although related to architectural exploration, are beyond the scope of this book.

An Introduction to Behavioral Synthesis

2.2.2 Exploring Alternate Architectures

Once the algorithm has been developed and verified, an architecture must be selected and described with an RTL-level description. This RTL-level description is typically disjoint from the original behavioral description, and its capture is largely a manual process, requiring significant time and effort in the context of the overall design cycle.

The process of selecting an architecture is a critical step in the design process, but one that rarely receives the attention it deserves. This is not the fault of the design teams, but largely results from shortened schedules that stem from competitive pressures. Until recently, there have been few, if any, design tools that enable the designer to quickly evaluate architectural alternatives. Without the use of such tools, the time required to complete the implementation of one particular architecture typically precludes the evaluation of alternate architectures.

Alternate architectures for an algorithm usually trade off area for latency (number of clock cycles to perform the calculation) as shown in Figure 2-5.

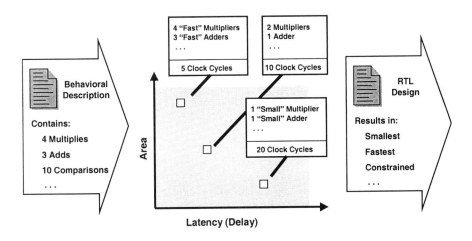

Figure 2-5: Architectural alternatives from a behavioral description

Once the RTL model is complete and verified, RTL synthesis is performed and the gate-level result is validated. If, at this point, the design is not within specification and the designer is unable to meet the design goals through successive iterations in RTL synthesis, portions of the architecture (and, possibly the entire architecture) must be re-addressed to solve the problem or problems. This can be a very expensive iteration.

> *If an architecture is found to be unacceptable in the final stages of the design process, the entire project schedule must be extended. This can be very expensive, even devastating, for products which are sensitive to time to market pressures.*

2.3 The Behavioral Design Process

Figure 2-6 shows a modified design process that utilizes behavioral synthesis.

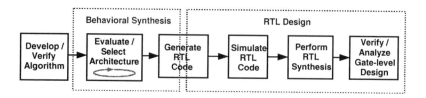

Figure 2-6: The design process utilizing behavioral synthesis

> *Behavioral synthesis automatically generates the RTL code for use in the traditional RTL design process.*

Behavioral synthesis allows designers to quickly evaluate many different architectures early in the design process. Because architectural decisions have such a major impact on the design schedule and the resulting implementation, this technology is receiving much attention from the design community.

Once the algorithm is complete, the behavioral specification itself is used as input to the behavioral synthesis tool. By varying design constraints (such as clock period, I/O timing, number and type of data path elements, and desired number of clock cycles), the designer can quickly evaluate of a number of architectures.

Once the behavioral synthesis tool has produced an architecture that meets the designer's specifications, the RTL code is then generated automatically by the behavioral synthesis tool.

> *Unlike in the RTL design flow, in the behavioral design flow there is a direct tie between the behavioral specification and the resultant RTL equivalent: the behavioral synthesis tool automatically generates the RTL code.*

The behavioral design flow offers productivity gains beyond the ability to generate multiple implementations for an algorithm. A more detailed comparison of the RTL design flow and the behavioral design flow is shown in Figure 2-7.

In the RTL design flow, a designer may write a behavioral model to validate the function of the algorithm. But for each architecture to be evaluated, the designer must write an RTL description and validate the functionality of that code. The entire RTL synthesis flow must be followed to determine if an architecture is satisfactory. If an architecture is not

An Introduction to Behavioral Synthesis

satisfactory, a new RTL model must be created and validated. This can lead to a long and possibly unpredictable process.

Unlike in the RTL design flow, in the behavioral design flow there is a direct tie between the behavioral code and the RTL code that is synthesized. In the behavioral flow, multiple architectures are quickly evaluated using the behavioral synthesis tool. When a satisfactory architecture is obtained, the behavioral synthesis tool automatically generates the RTL code. This tight coupling between the behavioral specification and the resulting RTL code greatly reduces, if not completely eliminates the possibility of having to reiterate on the gate-level results.

In the behavioral design flow, even if it is necessary to select a different architecture because the characteristics of the gate-level design are inadequate, this is a relatively simple process. The constraints supplied to the behavioral synthesis tool can be modified to create an alternate architecture – but the behavioral code need not be changed. In the RTL flow, new code must be developed to represent the architectural changes that were made to the design.

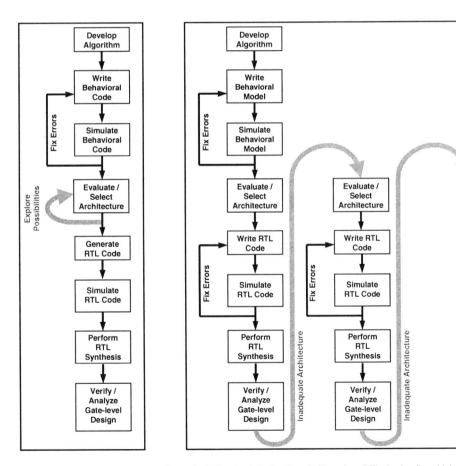

Figure 2-7: Detailed comparison of a behavioral design flow (left) and an RTL design flow (right)

2.3.1 A Simple Algorithm

For historic and development reasons many algorithms are described using "C" or a similar programming language. Consider the simple algorithm in Figure 2-8, captured in "C".

```
int design( int a, int b, int c, int d, int e, int f )
{
   int y;

   y = a * b + c + ( d - e ) * f;
   return y;
}
```

Figure 2-8: A simple "C" algorithm

For behavioral synthesis tools that do not support "c", this algorithm can be translated into VHDL so that it can be synthesized. The interface to the function can be represented as the design ENTITY and the body of the function as an ARCHITECTURE of the design, as shown in Figure 2-9.

```
ENTITY design IS
   PORT( a, b, c, d, e, f : IN  integer;
         y                : OUT integer );
END design;

ARCHITECTURE behavioral OF design IS
BEGIN
   PROCESS( a, b, c, d, e, f )
   BEGIN
      y <= a * b + c + ( d - e ) * f;
   END PROCESS;
END behavioral;
```

Figure 2-9: Behavioral VHDL

2.3.2 Data Dependencies

In order to understand the dependencies between data (a, b, c, d, e, f and y) and operations (*, +, -) in a design, it is often useful to represent the design as a *data-flow graph* (DFG).

> *A Data Flow Graph (DFG) depicts the inputs and outputs of a design, the operations used in the design, and the flow of data from the inputs to the outputs.*

Based on the precedence of operators as defined for the VHDL language, this design is shown as a DFG in Figure 2-10. The intermediate values have been labeled (*t1-t4*) so that they can be referred to in this discussion.

An Introduction to Behavioral Synthesis

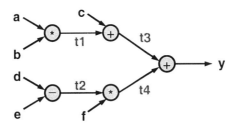

Figure 2-10: Data Flow Graph (DFG)

Each *node* in the graph represents an operation (such as a multiply or an add). Two nodes are connected if there is a data dependency between them. Data dependencies indicate the order in which values must be calculated. For example, the value **a * b** must be calculated before **t1 + c** can be calculated.

A DFG shows only data dependencies. The most parallel combinational implementation of an algorithm can be found if a unique resource (32-bit adder, 32-bit multiplier, etc.) is used to implement every node in a DFG. Data flow graphs are discussed in greater detail in Chapter 3.

2.3.3 RTL Synthesis Implementation

Consider how the design in Figure 2-9 would be synthesized using conventional RTL synthesis tools.

For RTL synthesis tools, a PROCESS is either "combinational" or "clocked". The process that describes this design is a combinational process. A combinational implementation of this process requires two multipliers, two adders, and one subtractor.

Because the interface signals are declared to be unconstrained integers (in the port declaration of the VHDL), they will get translated into 32-bit bus representations. The circuit in Figure 2-11 depicts the design when synthesized with an RTL synthesis tool.

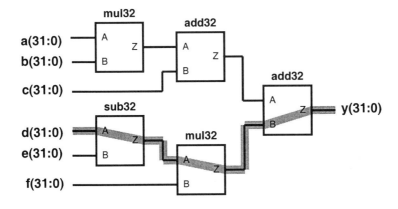

Figure 2-11: Combinational Implementation

The total area required for this implementation is shown in Figure 2-12.

$$\text{area}(\text{total}) = 2 * \text{area}(\text{mul32}) + \\ 2 * \text{area}(\text{add32}) + \\ 1 * \text{area}(\text{sub32})$$

Figure 2-12: Total area of combinational implementation

And the critical path (assuming all inputs arrive at the same time and the delay of a subtractor is greater than the delay of an adder) is shown in Figure 2-13 (see highlighted area in Figure 2-11).

$$\text{delay}(\text{total}) = \text{delay}(\text{sub32}) + \\ \text{delay}(\text{mul32}) + \\ \text{delay}(\text{add32})$$

Figure 2-13: Possible critical path of combinational implementation

2.3.4 Resource Allocation

Although this architecture is likely to be the *fastest* implementation (as everything is happening with maximum parallelism), it may not meet the design requirements. Multipliers with large bit widths require a lot of area. The designer may be willing to consider an alternate architecture that is slower but requires less area. This design contains two multiply operations and two add operations. An obvious architecture to investigate is one that requires only one instance of each of these operators.

> *Resource allocation is the process of deciding how many and which kind of resources can be used in a given implementation.*

An architecture in which resources are limited will have a very different structure from the previous combinational implementation. There are a number of implications associated with limiting resources. Consider the multiply operator: if only one multiplier is to be used in this architecture, then the single resource will be needed to multiply **a** and **b** as well as **t2** and **f** (see Figure 2-10). This means the two multiplication operations will have to occur at different times. The area is reduced at the cost of parallelism, and thus performance. The inputs to the multiplier must be multiplexed so that the appropriate input data is selected. In addition, the output of the multiplier will need to be registered so the result of the first multiply can be held stable while the second is being computed.

When any algorithm is implemented with limited resources, the use of those resources must be scheduled into separate clock cycles. If a calculation is performed in multiple clock cycles, the design will almost always contain a state machine. And if values

An Introduction to Behavioral Synthesis

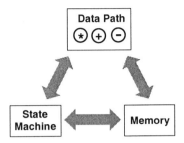

Figure 2-14: General architecture of scheduled designs

calculated in one clock cycle must be saved for use in another clock cycle, the design must contain memory (or registers).

At a very high level, a resource-limited implementation of a design implies the general architecture shown in Figure 2-14.

> Behavioral synthesis produces designs with a set number of data path resources (such as multipliers and adders). A state machine will be generated to control the use of these resources and intermediate values will be stored in some kind of memory or registers.

2.3.5 Scheduling

To implement a design using the general architecture of Figure 2-14, it is necessary to start with the Data Flow Graph, and decide when (in which clock cycle) the operations in the design will be performed.

For simplicity, assume that each operation will be implemented with components that each require an entire clock cycle to calculate a result. The DFG for this design can be modified to represent this additional information.

In Figure 2-15, each operation has been assigned to a clock cycle. Thus, three clock cycles are required to complete the desired calculation.

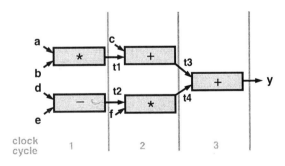

Figure 2-15: Modified DFG with scheduling information

Given a schedule, the three elements of the architecture illustrated in Figure 2-14 (state machine, memory, and data path) can be constructed.

2.3.6 Control Logic / State Machine

The state machine needed to control this design has three states. The state diagram for this machine, along with a fragment of VHDL code that illustrates how this state machine could be described for RTL synthesis, is shown in Figure 2-16. The states of the state machine are referred to as *s1*, *s2*, and *s3*. For clarity, the state diagram does not show the transitions for the synchronous reset.

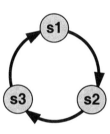

```
PROCESS( state )
   CASE state is
      WHEN s1 =>
         next_state <= s2;
      WHEN s2 =>
         next_state <= s3;
      WHEN state_3 =>
         next_state <= s1;
   END CASE;
END PROCESS;

PROCESS( clk )
BEGIN
   IF clk'EVENT AND clk = '1' THEN
      IF ( rst = '1' ) then
         state <= s1;
      ELSE
         state <= next_state;
      END IF;
   END IF;
END PROCESS;
```

Figure 2-16: Control state machine and RTL VHDL description

The VHDL code in Figure 2-16 is actually incomplete. The state machine will have to set the value of output control signals that will affect the multiplexing of data into arithmetic components and registers.

2.3.7 Memory

This architecture requires that certain intermediate values be somehow stored in memory. These values could be stored in RAM or (more likely for this simple design) in registers. A closer analysis of the design is necessary to determine which values must be registered.

Referring to Figure 2-15, it has already been noted that the outputs of the shared resources (the adder and multiplier) must be registered. The value of **y** is actually the same as the (registered) output of the adder. But the intermediate value **t2** is not so

An Introduction to Behavioral Synthesis

obvious. If the values of **d** and **e** are *only* valid during the first clock cycle, then **t2** must be registered for use in the second clock cycle. Note that the schedule in Figure 2-15 also implies that the input signals **c** and **f** are read in the *second* cycle, not the first.

If the input signals **c** and **f** were to be read in the first cycle, then they would have to be registered as well and held until their values were needed. A schedule in which all inputs are read in the first cycle is shown in Figure 2-17.

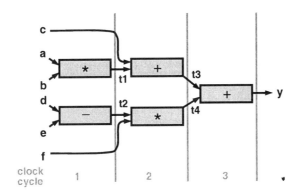

Figure 2-17: Modified schedule with all inputs read in the first cycle

> *If an arrow crosses a cycle boundary in a schedule, the data it carries must be stored in a memory element (RAM or register)*

Clearly, the I/O requirements of the design can have significant impact on the resulting implementation.

2.3.8 Data Path Elements

Working from the schedule from Figure 2-15, a determination can be made as to how the resources that appear in the schedule can be shared. Note that it is not always beneficial to share resources. Multiplexing logic that must be inserted in order to share resources make require greater area than an additional resource. Also, the additional delay through the multiplexing logic might result in an implementation that does not meet the timing requirements of the design.

After the resources to be shared have been identified, each element in the data path can now be combined with the appropriate multiplexing logic to control the inputs to that resource and with the necessary registers to store the values at the outputs of the resource. For example, Figure 2-18 shows the multiplexing and register required to use a single multiplier to perform the multiply operations that were scheduled in cycles 1 and 2. In this case, the area of the additional hardware required to use only a single multiplier is much smaller than the area of an additional multiplier.

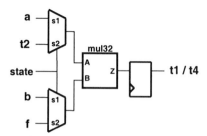

Figure 2-18: Multiplier with surrounding multiplexing logic and registering

When the state machine is in state *s1*, the values **a** and **b** will be multiplied together. When the state machine is in state *s2*, the values **t2** and **f** will be multiplied together. The result is stored in a register. This structure can be described with the VHDL code in Figure 2-19.

```
PROCESS( a, b, f, t2, state )
    IF ( state = s1 ) THEN
        mul32_a_in <= a;
        mul32_b_in <= b;
    ELSE
        mul32_a_in <= t2;
        mul32_b_in <= f;
    END IF;
END PROCESS;

PROCESS( clk )
BEGIN
    IF ( clk'EVENT AND clk = '1' ) THEN
        t1 <= mul_out;
    END IF;
END PROCESS;


mult: mul32
    PORT MAP( A => mul32_a_in,
              B => mul32_b_in,
              Z => mul32_out );
```

Figure 2-19: RTL VHDL to instantiate multiplier, multiplex inputs, and register output

Actually, the code in Figure 2-19 does not exactly represent the functionality of the schematic in Figure 2-18. The schematic does not describe the values of the inputs to the multiplexers for state **s3**. The VHDL code uses the same input values for state **s3** as are used in **s2**. This minor difference allows the VHDL code to be simpler.

Similar code could be constructed for the other data path elements in the design. Note that the output of the subtractor will be connected to a register, but the inputs are not multiplexed.

An Introduction to Behavioral Synthesis

2.3.9 Gantt Chart

The assumptions made to produce this implementation are overly simplistic. In particular, it was assumed that each operation could be performed in one clock cycle. But a 32-bit addition and a 32-bit multiplication can have very different delays. Forcing a clock cycle long enough for the multiplier leaves wasted time in the add cycles.

Now, assume that the designer has allocated the resources shown in Figure 2-20. These resources have actual delay numbers.

Resource	Quantity	Delay
Multiply	1	14 ns
Add	1	5 ns
Subtract	1	5 ns

Figure 2-20: Allocation example with limited resources

If the desired clock cycle is 25 ns, the design could be scheduled as shown in Figure 2-21. The schedule is shown as a *Gantt chart*.

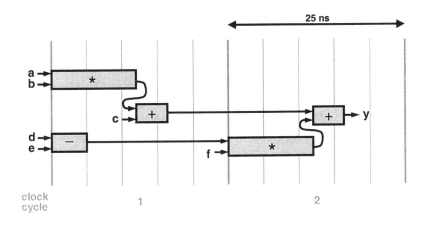

Figure 2-21: Scheduled design shown as a Gantt Chart

> *A Gantt chart, often used to display work schedules, shows the actual time required to perform each operation and the data dependencies between operations.*

The width displayed for each operation represents the propagation delay of the hardware component required to perform that operation.

2.3.10 Chained Operations

Based on a 25 ns clock period and the relative delay of the specified resources, the entire sequence of operations can be performed in just two clock cycles. Only one multiplier and one adder are required, as they are each used in different clock cycles. This schedule is different from the previous schedule in Figure 2-17 in that the result of the multiplication is *immediately* passed to an adder (the output of the multiplier is not registered).

> *When two data dependent operations are scheduled in the same clock period, the operations are said to be chained.*

Note that the combinational implementation is essentially achieved when there are sufficient resources and time to chain all the operations into a single clock cycle.

2.3.11 Multi-cycle operations

Consider another architecture for which the desired clock period is **10 ns**, but there are no restrictions on the *number* of resources that can be used. The allocation of resources (i.e. the total number and type *available*) is shown in Figure 2-22.

Resource	Quantity	Delay
Multiply	2	14 ns
Add	2	5 ns
Subtract	1	5 ns

Figure 2-22: Allocation example with unlimited resources

Even though the delay through a multiplier is actually longer than the clock period, it is still possible to schedule this design. The schedule is shown in Figure 2-23.

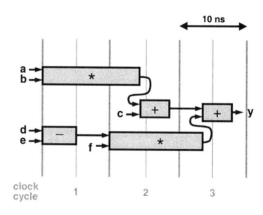

Figure 2-23: Gantt Chart showing multi-cycle operations

An Introduction to Behavioral Synthesis

> *An operation that is scheduled across two or more clock periods is a multi-cycle operation.*

Note that if the multiplier is combinational, the inputs to the multiplier must be held stable for two clock cycles to implement this schedule.

Even though it was assumed that unlimited resources were available in this example, only one adder component is actually required, since the add operations do not take place in the same cycle.

2.4 Summary

As demonstrated, even for this very simple design, there are several different possible architectures, each of which may address different design goals. For designs of considerably greater complexity, the evaluation of multiple architectures would be extremely difficult and time consuming if scheduling had to be performed manually.

Fortunately, behavioral synthesis tools perform these tasks automatically. Given a set of constraints specified by the designer (such as clock period, number and type of data path elements, and desired number of clock cycles), behavioral synthesis automatically schedules the operations into clock cycles and constructs the appropriate state machine to control those data path elements. Such tools can generate an RTL representation that can be processed using traditional RTL synthesis and gate-level optimization tools.

Chapter 3

The Behavioral Synthesis Process

3.1 The Behavioral Synthesis Process

This chapter discusses the steps performed during behavioral synthesis. While this chapter discusses some topics that are not essential to the *use* of behavioral synthesis tools, a good understanding of these steps will provide insight into the results produced by behavioral synthesis.

The general steps performed by a behavioral synthesis tool are shown in Figure 3-1. These steps are discussed in the sections that follow. Note that this flow shows a simple depiction of one possible implementation of a behavioral synthesis system. The ordering of steps in the flow may differ among systems. In addition, for simplicity, interactions that may occur between these blocks are not depicted.

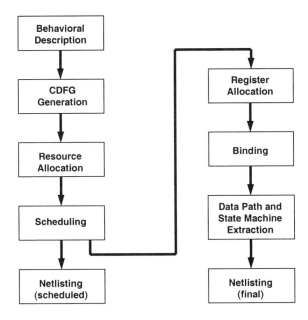

Figure 3-1: General steps in the behavioral synthesis process

3.2 Internal Representations

The first step in the behavioral synthesis process is the translation of the behavioral description into a Control / Data-Flow Graph (CDFG).

> *A Data Flow Graph (DFG) depicts the inputs and outputs of a design, the operations used in the design, and the flow of data from the inputs to the outputs. A Control / Data Flow Graph (CDFG) is an extension of a DFG that allows the representation of control structures, such as conditionals and loops.*

In a CDFG, each *node* in the graph represents an *operation* (such as a multiply or an add). Two nodes are connected if there is a data dependency between them. The connection between two nodes is called a *dependency arc*. Dependency arcs indicate the order in which operations must be performed. For example, in the expression **((a + b) * c)**, the addition must be completed before the multiplication can begin.

An example of a CDFG is shown in Figure 3-2.

```
IF ( n > m ) THEN
    tmp := a + b;
ELSE
    tmp := ( a * b ) + c - d;
END IF;
y <= tmp + e;
```

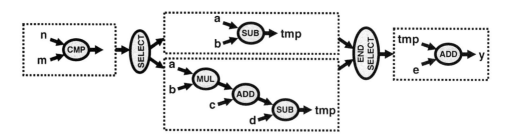

Figure 3-2: Example of a Control / Data Flow Graph (CDFG)

The VHDL code in the example includes an IF statement. Control operations such as this are represented in the CDFG using special nodes. In this diagram, the nodes that are labeled "SELECT" and "END SELECT" represent the start and end of the conditional, respectively. Some synthesis tools use the terms "FORK" and "JOIN" to represent these concepts.

The Behavioral Synthesis Process

Similar terms are used for looping constructs. For example, "LOOP BEGIN" and "LOOP END" nodes represent the start and end of a loop, respectively. NEXT and EXIT statements are represented by "ITERATE" and "TERMINATE" nodes.

A CDFG represents the functionality of a design. But the CDFG does not indicate the resources that will be used to implement that functionality, nor the number of clock cycles required to perform the calculation. This information is not known until after scheduling.

> *Data dependencies, which are represented in the CDFG, cannot be violated during the synthesis process. This information, along with a set of resources and I/O timing constraints, is used during the scheduling process to generate scheduled designs.*

3.3 Resource Allocation

One of the first steps in the behavioral design process is resource allocation. The concept of resource allocation was first introduced in Chapter 2.

> *Resource allocation is the process of deciding how many and which kind of resources can be used in a given implementation.*

The number and type of allocated resources can significantly impact the results of synthesis. As shown in Chapter 2, there is a direct correlation between allocated resources and performance. A more parallel implementation can be achieved when a large number of resources have been allocated. When only a small number of resources are available, those resources must be reused during the calculation of the algorithm, resulting in a longer schedule.

A CDFG shows only data dependencies. It does not indicate the resources that will be used to implement that functionality. Resource allocation merely defines the set of components that *could* be used to implement that functionality.

The relationship between a component and the operations in a CDFG is defined by the *component bindings* that are defined for that component. A component binding defines how a component can be used to implement the functionality of an operator.

Consider a simple algorithm and its associated DFG shown in Figure 3-3. The DFG contains three operators: ADD, SUB, and MUL. These operators were inferred from HDL code.

$$y = ((a*b)+c)+((d-e)*f)$$

Figure 3-3: Simple algorithm and associated DFG

For simplicity, assume that all operations are 16 bits. In order to schedule this design, it is necessary to allocate resources that have component bindings for a 16-bit add operation, a 16-bit multiply operation, and a 16-bit subtract operation.

Figure 3-4 illustrates the possible bindings of operators to components that might be defined for a particular technology library.

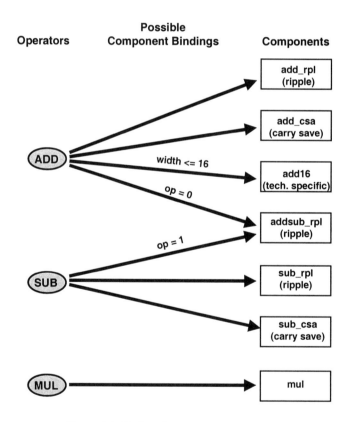

Figure 3-4: Binding of operators to components

The Behavioral Synthesis Process

Note that component bindings are technology-specific. For example, Figure 3-4 shows that the technology has a hard macro (**add16**) that can perform a 16-bit add. This component is only available when targeting the particular technology. In addition, the component can only be used if the width of the inputs is less than or equal to 16.

The figure also shows how there is a choice about which component a particular operation is bound to -- an ADD operation can be performed using a ripple adder, a carry-save adder, or a hard-macro adder. Permissible component bindings can be dependent upon properties of the operator, as is the case for the hard macro, which is limited to a width of 16 (or fewer) bits.

Component bindings can also specify how the signals of a component should be controlled. For example, both an ADD operation and a SUB operation can be bound to an **addsub_rpl** component, but the **op** control line must be set to zero or one, depending upon the operator being bound.

For any given design, every operation in the CDFG must have at least one possible binding to an allocated resource. In the previous example, it is necessary to allocate *some* resource that can be used to perform a 16-bit multiplication.

The bindings in Figure 3-4 show how a small number of resources could be used to implement this design. For example, a single 16-bit **addsub_rpl** component could be used to implement all of the add and subtract operations. In addition, a multiplier component would be required to implement the multiply operations.

> *A particular resource allocation is considered sufficient, if it is possible to implement every operation in the CDFG with one or more of the allocated resources. A design with an insufficient allocation of resources cannot be scheduled.*

In a large design, it is often important to keep the number of large components (such as multipliers) to a minimum. Even if a design contains a large number of multiply operations with different bit widths, it is not necessary to allocate a separate multiplier for every bit width. A sufficient resource allocation can be achieved by only allocating a small number of large multipliers.

For example, a design that contains a 32-bit multiplication, a 24-bit multiplication, and a 16-bit multiplication can be implemented using a single 32-bit multiplier. For the 24-bit and 16-bit multiply operations, the upper bits of the 32-bit multiplier can be connected to ground or signed extended (depending on the operation) to produce the correct results. Behavioral synthesis tools understand how one operation "covers" another operation. When a component is used in this manner, the behavioral synthesis tool will appropriately connect the inputs and outputs to correctly implement the functionality.

Resource allocation can be an automatic or a manual process. For example, the designer may specify that the design be scheduled with the minimum number of resources, allowing the synthesis tool to select a sufficient allocation based on the library's component bindings. Alternately, the designer can hand-allocate the resources that can be used in a particular implementation.

3.4 Scheduling

Scheduling is the enabling technology of behavioral synthesis. The quality of results produced by a behavioral synthesis tool is most influenced by the quality of the scheduling algorithms employed by the tool.

Commercial behavioral synthesis tools are likely to use proprietary scheduling algorithms. But it is also likely that these algorithms are related to those discussed in this section.

Papers that discuss scheduling algorithms have been published in academic publications since about 1990. Although commercial behavioral synthesis tools are relatively new, the discussion of scheduling itself is not.

This section discusses some of the more common scheduling techniques, to provide some insight into this part of the behavioral synthesis process.

3.4.1 Scheduling Concepts

Scheduling algorithms can be grouped into general categories: those that produce schedules for unconstrained designs and those that produce schedules for constrained designs.

Scheduling algorithms designed for unconstrained designs can be used to quickly produce feasible schedules that are restricted only by data dependencies. While these algorithms alone rarely address real-world problems, the results they produce can be used as a measure against the results of other algorithms. As Soon As Possible (ASAP) and As Late As Possible (ALAP) scheduling can be used to address the unconstrained scheduling problem.

A design can be resource-constrained, time-constrained, or both. For a resource-constrained design, the goal of scheduling is to produce the fastest implementation (i.e. minimize latency) using the resources provided. For a time-constrained design, the goal of scheduling is to produce the smallest implementation without violating the timing constraints. List scheduling and force-directed scheduling are techniques that were developed to address the resource-constrained and time-constrained scheduling problems, respectively.

3.4.2 ASAP / ALAP Scheduling

The ASAP and ALAP scheduling algorithms provide a simple and fast solution to the unconstrained scheduling problem. As the name implies, with ASAP scheduling, operations are scheduled in the earliest possible clock cycle. Similarly, with ALAP scheduling, operations are scheduled into the latest possible clock cycle. A simple algorithm, ASAP, and ALAP schedules are shown in Figure 3-5.

While these algorithms rarely address real-world problems by themselves, they are often used as a part of other, more compute-intensive algorithms to help determine the bounds of the problem.

The Behavioral Synthesis Process

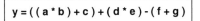

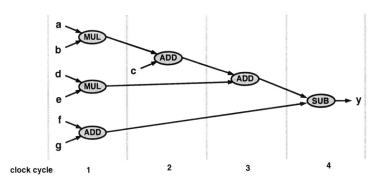

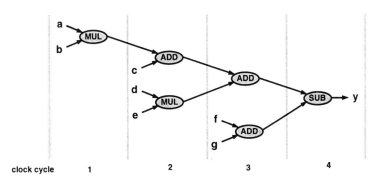

Figure 3-5: Simple algorithm scheduled using ASAP and ALAP scheduling algorithms

3.4.3 Other Scheduling Methods

List scheduling is a scheduling algorithm that can be used to schedule a resource-constrained design. The algorithm begins at the first clock cycle and continues until all operations have been scheduled. Operations to be scheduled are selected from a *ready list*, which includes every operation that could be scheduled in the current clock cycle without violating data dependencies. The key to this algorithm is that the ready list is a *prioritized* list -- operations in the list are ordered according to a heuristic. For example, some implementations of this algorithm order the operations by mobility.

> *The mobility of an operation is defined by the range of clock cycles in which the operation could be scheduled without changing the overall schedule length.*

The operations are ordered by mobility under the assumption that operations with lesser mobility are more difficult to schedule. There has been much research about the heuristic that produces superior results in the majority of cases.

Figure 3-6 shows an ASAP schedule of the simple algorithm. The mobility of the operations is also shown.

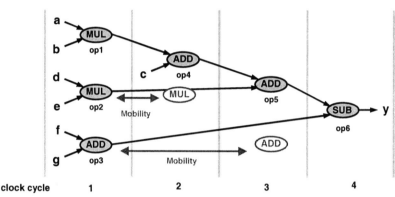

Figure 3-6: ASAP schedule of a simple algorithm showing the mobility of operations

Figure 3-7 shows the development of a schedule using list scheduling, assuming the minimal number of components. The priority function is based on the mobility of the operations.

For each clock cycle, a ready list is created. This list includes all of the operations that *could* be scheduled in the clock cycle. The list is prioritized according to mobility. Less mobile operations have a higher priority. Beginning at the top of the list, operations are scheduled until all of the available resources have been used. This example assumes that only 1 multiplier, 1 adder, and 1 subtractor are available. The process continues with the next clock cycle, until all operations have been scheduled.

Force-directed scheduling is a more complicated scheduling algorithm that can be used to address the time-constrained scheduling problem. The algorithm begins by constructing ASAP and ALAP schedules, taking timing constraints into account. This results in the determination of a mobility window for each operation in the schedule. Without delving into the details of the algorithm (which are beyond the scope of this book), this information is used to determine the "force" (i.e. area cost) associated with scheduling each operation into a particular clock cycle within its mobility window. This force includes the "indirect" cost contributed by other operations that would be effected by that decision. This global analysis produces a uniform distribution of operations across the schedule.

The Behavioral Synthesis Process   33

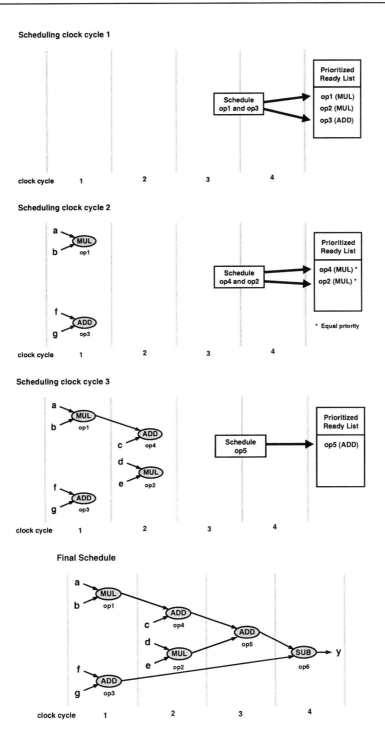

Figure 3-7: Development of a schedule using list scheduling

For completeness, it is appropriate to mention that integer linear programming (ILP) techniques can be used to address the scheduling problem. Unlike the previous algorithms that are based on heuristics, integer linear programming can provide optimal solutions to the time and resource-constrained scheduling problems. Unfortunately, these techniques are so computationally intensive, they are usually unusable for any real-life scheduling problems.

There has been much research devoted to the scheduling problem. This section simply highlights a few easily-understood approaches. Scheduling algorithms that address actual design problems are generally much more complicated than those described here. Some factors that complicate the scheduling process are:

- Chained operations

    When two or more data dependent operations are scheduled in the same clock period, the operations are chained. Chaining can be used to decrease latency. This may come at the cost of an increased number of resources (chained operations are active concurrently) or at the cost of larger resources (they must be sufficiently fast to fit with a clock period).

- Multi-cycle operations

    An operation that is scheduled across two or more clock periods is a multi-cycle operation. An operation, such as a multiplier, might have bindings to fast (single-cycle) components and small (multi-cycle) components. To effectively utilize these components, the trade-off between area and latency must be made during scheduling.

- Multiple possible bindings

    For any particular operation, a technology library will typically define many possible bindings. For example, an add operation could be implemented with a ripple adder or a carry-lookahead adder. It may be necessary to use faster components on the critical path of a design, but use smaller components for the rest of the design to save area.

Many of the books listed in the References and Resources portion of this book provide a more complete discussion of scheduling algorithms. This discussion of algorithms is intended to provide insight into the general approaches to performing scheduling.

The Behavioral Synthesis Process

3.5 Register Allocation

As discussed in Chapter 2, if a dependency arc crosses a cycle boundary after scheduling, the data it carries must be stored in memory or a register. But this does not mean that a unique register is implied for every such crossing.

It is possible to reduce the number of registers that are required to implement a scheduled design by performing *lifetime analysis* on the data that must be stored in registers and then by sharing registers whose lifetimes do not overlap.

3.5.1 Lifetime Analysis

Lifetime analysis is used to determine the clock cycles in which data is valid. This information can be used to share registers in the implementation of a scheduled design.

Consider the very simple design and schedule in Figure 3-8.

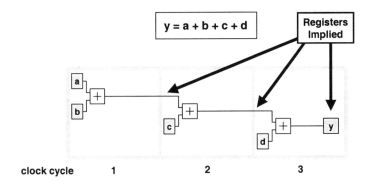

Figure 3-8: Simple algorithm scheduled with a single adder component

The design contains 3 add operations, and has been scheduled assuming only a single adder component. This results in a schedule that is 3 clock cycles long.

The data represented by each dependency arc that crosses a cycle boundary must be stored in a register. In addition, all outputs of synthesized designs are registered. Assuming that each add operation is 8 bits wide, it appears that the implementation of this schedule requires 3, 8-bit registers.

But the data represented by the dependency arcs are not valid for *every* clock cycle in the schedule. For example, the output of the first add (in clock cycle 1) must be registered so that the value can be used in clock cycle 2. But that data is not used at all in clock cycle 3. The data represented by that dependency arc has a *lifetime* that begins at the start of clock cycle 2 and ends at the end of the same clock cycle -- it is only necessary that the data be saved for that one clock cycle.

Similarly, the output of the second add (in clock cycle 2) has a lifetime that begins at the start of clock cycle 3 and ends at the end of the same clock cycle – this data is also only needed for that one clock cycle.

Recall from Chapter 2, the output of synthesized designs are registered. Thus, the lifetime of the data associated with the output signal **y** is much longer than the lifetime of the other signals. This value is only updated when a new output value is calculated. Thus, the lifetime of this data is the entire schedule.

The lifetimes of the data that must be stored for this schedule are shown in Figure 3-9.

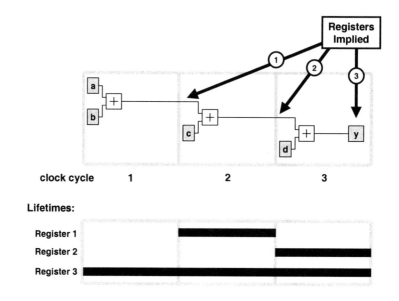

Figure 3-9: Scheduled design and associated lifetime analysis

3.5.2 Register Sharing

Lifetime analysis can be used to share registers. Figure 3-9 shows that the lifetimes of register 1 and register 2 are *non-overlapping*. This means that the same registers can be used to hold both values. There is no time in the schedule when both pieces of data are valid.

If these two registers are shared, the implementation of this schedule requires only 2, 8-bit registers. Because the lifetime of register 3 is the entire schedule, it cannot be shared with any other registers.

> *Registers with non-overlapping lifetimes can be shared. Register sharing can significantly reduce the size of the final design.*

3.6 Binding

Binding is the process of assigning each scheduled operation to a component. Allocation ensures that there are sufficient resources to implement the design and scheduling uses those resources to assign operations to clock cycles. But the actual components to be used for each operation are not assigned at that time.

Consider the simple algorithm and associated schedule in Figure 3-10. Two adder components and one multiplier component were allocated to produce the schedule.

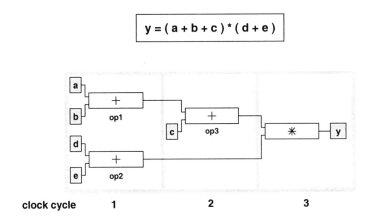

Figure 3-10: Simple algorithm and associated schedule

The binding of the multiply operation is simple. The one multiply operation is bound to the one multiplier component.

There are two possible bindings for the adder components. There are three add operations, labeled **op1**, **op2**, and **op3** in Figure 3-9. The operations **op1** and **op2** must be bound to different adder components, as they are used concurrently. The operator **op3**, however, could be bound to either adder component. These possible bindings are shown in Figure 3-11.

Binding 1

Operation	Binding
op1	add1
op2	add2
op3	add1

Binding 2

Operation	Binding
op1	add1
op2	add2
op3	add2

Figure 3-11: Possible bindings for the add operations

It may seem that these two bindings will yield equivalent implementations. But binding choices can impact the amount of multiplexing and interconnect wiring required in the final implementation.

This is illustrated by the implementations of the two different bindings for this design. The two implementations are shown in Figure 3-12.

Binding 1

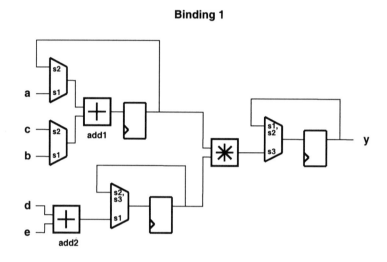

Binding 2

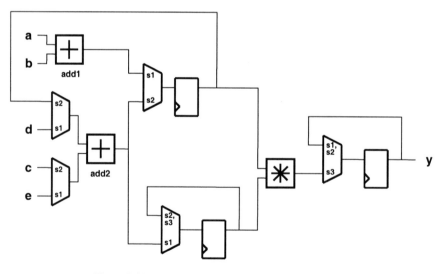

Figure 3-12: Implementations for the two possible bindings

For this simple design, the first binding requires one less multiplexer than the second binding. For a large design, the choice of bindings can save significant area in the final implementation.

> *Binding choices can impact the amount of multiplexing and interconnect wiring required in the final implementation.*

3.7 Data Path and State Machine Extraction

After all operations have been bound to components, the data path and state machine can be constructed.

The data path consists of the components to which operators have been bound, the multiplexing logic that controls the inputs to and outputs from those components, as well as the registers that are required to store intermediate values.

The select lines of the multiplexers pass data to the components in the data path based on the current state of the state machine. The state machine is extracted from the schedule.

3.8 Netlisting

A netlist is used to pass design data from one tool to another. In the case of behavioral synthesis, a final netlist of the synthesized design is processed using the traditional RTL design process.

Some behavioral synthesis tools can generate a netlist after scheduling, but before binding and data path / state machine generation. Such a netlist can be used to validate the schedule of the design in the context of the surrounding system early in the design process.

3.8.1 Scheduled Design

The use of behavioral synthesis provides benefit to the *entire* design process. By describing functionality at the behavioral level, system-level simulation becomes more feasible. Because a behavioral description is smaller and contains fewer communicating signals than an equivalent RTL description, simulating a behavioral description requires significantly less time and fewer resources than simulating an equivalent RTL description.

Behavioral synthesis has the ability to quickly generate alternate architectures with potentially different I/O timing and latency. After a schedule has been generated, it is useful to be able to validate the timing and latency of the scheduled design in the system context.

Some behavioral synthesis tools provide this feature. The netlist produced after scheduling is *cycle-accurate,* which means that the cycle-to-cycle behavior of the netlist will be identical to the final design. However, operators are netlisted in an abstract manner (for example, +, instead of a collection of gates) to still facilitate extremely fast simulation.

The simulation performance of a cycle-accurate design does not equal that of a behavioral description, but is still significantly faster than the simulation of the equivalent RTL or gate-level models.

The relative simulation times for the JPEG compression algorithm (discussed in Chapter 14) are shown in Figure 3-13.

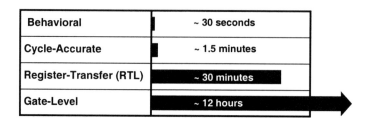

Figure 3-13: Simulation time comparisons

The use of a cycle-accurate scheduled netlist provides a huge benefit to the system verification aspects of a design effort.

3.8.2 Final Design

The final netlist produced by behavioral synthesis is often referred to as an "RTL" netlist. A better term for this netlist would be a "Structured RTL" netlist.

After binding, data path generation, and state machine generation, the design consists of data path components, multiplexer, registers, and a state machine.

The data path portion of the design is composed of a series of data path and register component instantiations, connected by instantiated multiplexers. This portion of the design is entirely structural.

The state machine portion of the design can be represented as an RTL description. By netlisting at this level of abstraction, the designer can still control the manner in which the state machine is implemented in gates (such as state encoding).

3.9 Summary

The goal of any tool is to shield the user from its underlying complexities. A designer must be able to interact with even the most complex synthesis tool in an easy-to-understand manner. At the same time, a designer is less likely to adopt a tool based on new technology without some understanding of its inner workings.

Understanding the basic terms associated with behavioral synthesis provide insight into the intermediate results produced by behavioral synthesis tools and helps a designer to guide the architectural exploration process.

Chapter 4

Data Types

4.1 Synthesis Considerations

A number of different data types can be expressed using the VHDL language. A designer must understand which data types are supported by behavioral synthesis and how those data types are translated into a gate-level netlist.

> *Since a gate-level netlist consists of gates and wires, the job of any synthesis tool (be it behavioral or RTL) is ultimately to translate all data types to those data types which can represent wires.*

When considering hardware design, data types can be grouped into two major categories: non-abstract types that readily translate to hardware (e.g. *std_logic*) and more abstract types that have the appearance of software programming constructs (e.g. records).

The data types that are supported by behavioral synthesis tools are very similar to those supported by RTL synthesis tools. This is one area where RTL synthesis already permits a quite high level of abstraction. However, even though a large number of data types are supported by both technologies, there are important factors to consider when selecting data types to be used in a design.

> *In VHDL, the std_logic type is probably the most standard representation of a wire. This type and the composite (array) types built from std_logic (e.g. std_logic_vector, signed, unsigned), are the only types that are likely to be found in designs generated using synthesis.*

When VHDL is being used to describe hardware, it may seem that types should be limited to the non-abstract data types that are easily mapped to hardware, such as *std_logic* and *std_logic_vector*. There are three main reasons to consider other data types: construction, analysis, and flexibility.

- Construction

    The use of abstract data types can simplify the coding of a design and its future maintenance. For example, if in a particular algorithm individual elements of data are always grouped together, it is more understandable and probably requires less code to represent the grouped data using a RECORD type. This use of an abstract type is consistent with the goals of behavioral synthesis; designers can specify the desired function of a design at a more abstract level which is closer to the algorithm itself, rather than describe a single architecture which implements that algorithm.

- Analysis

    Abstract data types can also simplify analysis. Consider a design that has different operating modes, such as "sleep", "normal", and "burst". The mode might be stored in a variable in the design. Ultimately, these modes must be represented as bit patterns (on wires), such as "00", "01", "10". When debugging the design in a simulator, it is much more useful to view the text that represents the states ("sleep", "normal", "burst") than it is to view bit patterns ("00", "01", "10"). In this example, an *enumerated* type could be used to represent the modes.

- Flexibility

    A synthesis tool must translate abstract types into non-abstract types. Often, this can be accomplished in more than one manner. If a design is described using abstract types, often the decisions about how to map to "wires" can be controlled when synthesis is performed. For example, an enumerated type which has possible values of "sleep", "normal", and "burst" could be mapped into the bit patterns "00", "01", and "10" (i.e. binary encoding). Alternately, the values could be mapped into the bit patterns "001", "010", and "100" (i.e. one-hot encoding). This flexibility is only available if the design was originally written using the abstract enumerated type.

4.2 bit / bit_vector Types

The *bit* type is an overly simplistic representation of a wire. It is declared in the standard package as an enumerated type with two possible values: '0' and '1'. There is no way to represent an unknown (or uninitialized) state with a *bit* nor can high-impedance values be represented. In VHDL, enumerated types are initialized to their left-most value in the absence of an explicit specification of an initial value. This means that a design that is described using the *bit* type will initialize to all zeros by default (even if no reset sequence has been performed). This is a dangerous assumption for simulation.

> *Because the std_logic and std_logic_vector types can easily be used to represent a wire and offer superior functionality, bit and bit_vector types should be avoided.*

Data Types

4.3 boolean Type

The *boolean* type is an abstract type since it is modified by the synthesis process. However, the mapping from a *boolean* to a *std_logic* type is certainly simple and easy to understand. Since types are initialized to their left-most value, variables and signals of type *boolean* will initialize to false, and like *bits*, have only two possible values. This is a shortcoming of the *boolean* type. The designer should take care to initialize boolean values before they are used. If such an initialization is absent, it will not be detected during behavioral simulation.

> *The value of having boolean flags in a simulation may outweigh its shortcomings. If used in a design, a boolean type should be thought of as an abstract type and thus should not be used to directly represent wires.*

4.4 std_logic / std_logic_vector / signed / unsigned Types

The *std_logic* type is probably the most standard representation of a wire. This type and the composite (array) types built from *std_logic* (e.g. *std_logic_vector, signed, unsigned*), are the only types that are likely to be found in designs generated using synthesis.

For simulation, the two logic values defined for *bit* are not sufficient to model a wire in hardware. The IEEE 1164 standard defines the *std_logic* logic type as a 9-value enumeration with the values '0', '1', 'L', 'H', 'Z', 'U', 'X', 'W', and '-'.

The interpretation for synthesis of the values '0' and '1' is quite understandable. The values 'L' and 'H' are treated synonymously as '0' and '1', respectively. 'Z' is the high impedance value for a tri-state signal. The values 'U', 'X', 'W' and '-' are not as well defined for synthesis.

The *std_logic* type is defined in the package *std_logic_1164* package in the *ieee* library. The type can be used in a design by including the following statements at the top of the VHDL file in which they will be used:

```
LIBRARY ieee;
USE ieee.std_logic_1164.ALL;
```

The *std_logic_1164* package also defines the type *std_ulogic*, which is the *unresolved* equivalent of *std_logic*. This means that a signal of type *std_ulogic* can have only one driver; a signal of type *std_logic* can have many drivers.

The *std_ulogic* type is useful because it helps prevent a common VHDL coding error: assigning to a signal in more than one PROCESS. A design in which there are multiple assignments to a signal of an unresolved type (such as *std_ulogic*) will not compile.

> *The type std_logic models a wire more accurately than std_ulogic because it allows multiple drivers and thus can behave like a bus. The std_logic type has become the de facto "standard" representation of a wire.*

VHDL is a strongly typed language. For any type, the operators (such as AND, OR, NOT, XOR, +, -, *, /, etc.) must be defined. In VHDL, almost nothing comes "predefined". Consequently, the *std_logic* type and its vector equivalent, *std_logic_vector*, are not very useful without operator definitions. The basic logical operators (such as AND, NOT, OR, etc.) are defined for *std_logic* and *std_logic_vector* in the *std_logic_1164* package.

The *std_logic_vector* type requires a more complete set of operators (such as arithmetic and comparison operators) to be useful for real designs. But a *std_logic_vector* could be interpreted as either a signed value or an unsigned value. Many of the operator definitions will need to differ, depending upon this interpretation. For example, the hardware that is needed to determine if one signed number is greater than another signed number is very different from the hardware needed to compare two unsigned numbers.

> *In the early days of RTL synthesis, Synopsys developed two packages, std_logic_signed and std_logic_unsigned, which define a set of arithmetic, conversion, and comparison functions for std_logic_vector. These packages, located in the ieee library, interpret std_logic_vector as signed and unsigned values, respectively.*

To interpret the *std_logic_vector* type as representing signed values, the VHDL code would need to begin with the following declarations:

```
LIBRARY ieee;
USE ieee.std_logic_1164.ALL;
USE ieee.std_logic_signed.ALL;
```

Even though the *std_logic_signed* and *std_logic_unsigned* packages allow *std_logic_vector* to be interpreted as either signed or unsigned, many designs will contain both signed and unsigned values. If a design requires both interpretations, the packages *std_logic_signed* and *std_logic_unsigned* will not be sufficient. Synopsys developed another package, *std_logic_arith*, to address this issue. This package defines two types, *signed* and *unsigned*, which are both vectors of *std_logic* elements. By defining these two new types, all of the appropriate operators can be defined. The arithmetic, conversion, and comparison functions for the *signed* and *unsigned* types are also defined in the *std_logic_arith* package.

Data Types

> *When signed and / or unsigned arithmetic is needed in a design, it is simpler and more understandable to use the types signed and unsigned as defined in the std_logic_arith package. Avoid using std_logic_vector for signed or unsigned values.*

The *std_logic_signed*, *std_logic_unsigned*, and *std_logic_arith* packages define a number of conversion functions to translate from one type to another. At least one of these functions is used in virtually any design that uses one of these packages. Each conversion function and the package in which it is declared is listed in the table in Figure 4-1.

Pkg	Input Type	Output Type	Interface
std_logic_arith	unsigned	integer	conv_integer(arg: unsigned)
	signed	integer	conv_integer(arg: signed)
	integer	unsigned	conv_unsigned(arg: integer; size: integer)
	unsigned	unsigned	conv_unsigned(arg: unsigned; size: integer)
	signed	unsigned	conv_unsigned(arg: signed; size: integer)
	integer	signed	conv_signed(arg: integer; size: integer)
	unsigned	signed	conv_signed(arg: unsigned; size: integer)
	signed	signed	conv_signed(arg: signed; size: integer)
	integer	std_logic_vector	conv_std_logic_vector(arg: integer; size: integer)
	unsigned	std_logic_vector	conv_std_logic_vector(arg: unsigned; size: integer)
	signed	std_logic_vector	conv_std_logic_vector(arg: signed; size: integer)
std_logic_signed	std_logic_vector	integer	conv_integer(arg: std_logic_vector)
std_logic_unsigned	std_logic_vector	integer	conv_integer(arg: std_logic_vector)

Figure 4-1: Useful conversion functions for *std_logic_vector*, *signed*, and *unsigned*

> *Conversion functions can be used freely within behavioral descriptions. Even though the code of some conversion functions may appear complicated, these functions are translated into wires during synthesis.*

The logical operators (AND, OR, NOT, XOR, etc.) are not defined for the types *signed* and *unsigned*. But it is easy to transform *signed* or *unsigned* to *std_logic_vector*. There is no conversion function to perform this transformation, but a *signed* or *unsigned* type can be easily changed to a *std_logic_vector* with a *type conversion*. A type conversion is similar to a *type cast* in other programming languages and can be used to change one array type to another array type when the elements in each array are the same. In the case of *signed*, *unsigned*, and *std_logic_vector*, each element of the arrays have the same type: *std_logic*.

The general form of a type conversion is:

target_type(*expression*)

So to perform a logical AND operation on two *unsigned* variables and assign the result to another *unsigned* variable, a user could write:

```
var3 := unsigned( std_logic_vector( var1 ) AND
                  std_logic_vector( var2 ) );
```

The IEEE Design Automation Standards Committee (DASC) Synthesis Working Group developed a package that is similar to the *std_logic_arith* package, called *numeric_std*. This working group was formed to define a more "standard" package that was controlled by a group not affiliated with one particular company and would be more likely to be supported by all synthesis tools.

> *Most synthesis tools now support the numeric_std package, which is part of the IEEE 1076.3 standard. This package defines signed and unsigned types, and provides functionality similar to the std_logic_arith package.*

The *numeric_std* package addresses some of the limitations of the *std_logic_arith* package. The *std_logic_arith* package overloads implicit function declarations, which when used, creates ambiguous function calls. However, most simulation and synthesis tools provide mechanisms for working around this issue.

The *numeric_std* package provides a set of conversion functions similar to the *std_logic_arith* package. These functions are shown in Figure 4-2.

Note that the *numeric_std* package has significantly fewer conversion functions than the *std_logic_arith* package. When using the *numeric_std* package, many type conversions must be done using the type conversion capability of the VHDL language.

Data Types

Pkg	Input Type	Output Type	Interface
numeric_std	unsigned	integer	to_integer(arg: unsigned)
	signed	integer	to_integer(arg: signed)
	natural	unsigned	to_unsigned(arg, size: natural)
	unsigned	unsigned	resize(arg: unsigned; size: natural)
	integer	signed	to_signed(arg: integer; size: natural)
	signed	signed	resize(arg: signed; size: natural)

Figure 4-2: Useful conversion functions for *std_logic_vector*, *signed*, and *unsigned*

4.5 Integer Type

Integers are one of the most commonly used abstract types. Integers can be constrained to a particular set of values or can be left unconstrained. An unconstrained integer will be synthesized as 32-bits. Constrained integers will be mapped to the minimum number of bits required to represent the specified range. Positive ranges will be mapped into unsigned representations; ranges that include negative values will be mapped into a 2's complement signed representation. Examples of type declarations and resultant bit representations are shown in Figure 4-3.

Integer Range Declaration	Number of Bits	Representation
integer RANGE 0 TO 7	3	unsigned
integer RANGE -1 TO 3	3	signed (2s complement)
integer RANGE 0 TO 11	4	unsigned
integer RANGE -8 TO 1	4	signed (2s complement)
integer	32	signed (2s complement)

Figure 4-3: Translation of integer types to bits during synthesis

> *Always constrain the range of integers to the values that are valid for that signal or variable. This will ensure that no unnecessarily large hardware is generated. VHDL simulators perform range checking which will catch "out-of-range" errors during behavioral simulation.*

Synthesis tools perform any necessary padding or sign extension when assignments are made between integers of differing ranges.

4.6 Enumerated Type

Consider a design that has different operating modes, such as "sleep", "normal", and "burst". Ultimately, these modes must be represented as bit patterns (on wires), such as "00", "01", "10". When viewing waveforms in a simulator, it is much more useful to see the text that represents the states ("sleep", "normal", "burst") than it is to view bit patterns ("00", "01", "10").

An enumerated type is an abstract type with a discrete set of values. In this example, an enumerated type could be used to represent the three modes.

> *An enumerated type is an abstract type with a discrete set of values. When an enumerated type is synthesized, a unique bit pattern is assigned to each possible value for the enumerated type. By default, bit patterns are assigned to values using a sequential binary encoding scheme.*

The number of bits required to represent an enumerated type is determined from the number of values the type can represent and the encoding scheme used by the synthesis tool. An example of an enumerated type declaration and associated bit patterns (using a sequential binary encoding scheme) is shown in Figure 4-4.

```
TYPE mode_types IS ( sleep, normal, burst, idle, maintenance );
```

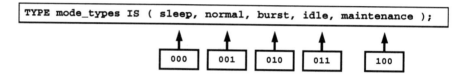

| 000 | 001 | 010 | 011 | 100 |

Figure 4-4: Translation of an enumerated type to bits during synthesis

Both RTL and behavioral synthesis tools allow the encoding for an enumeration to be controlled directly via attributes or commands. The attribute *enum_encoding* is a string that can be specified as shown in Figure 4-5.

```
TYPE mode_types IS (sleep, normal, burst, idle, maintenance);
ATTRIBUTE enum_encoding : string;
ATTRIBUTE enum_encoding OF mode_types : TYPE IS
    "00001 00010 00100 01000 10000";
```

Figure 4-5: Attribute to specify a particular encoding for an enumerated type

If sequential encoding is known to result in inadequate logic, an attribute can be specified. However, sequential encoding is usually satisfactory. Remember that one of the goals of working at the behavioral level is to rise above the bit-level details.

Data Types   49

Enumerated types should always be initialized. If they are not initialized, it is possible to start up with an illegal value. In the example in Figure 4-5, if a variable of this type is registered, and the register happens to get the value "11111" at power-up (the power-up state of registers is not predictable in most silicon technologies), the circuit may not work correctly.

4.7 Record Type

Records can be used to group data together. Consider an Asynchronous Transfer Mode (ATM) cell header. This 5-byte header consists of the following pieces of data:

Generic Flow Control (GFC): 4-bits

Routing Field (RF): 24-bits

Payload Type (PT): 3-bits

Cell Loss Priority (CLP): 1-bit

Header Error Check (HEC): 8-bits

To allow this group of data to be manipulated in a behavioral description as a single piece of data, a record type could be defined to represent the header.

A record type declaration defines the fields of data that comprise the record. Either an entire record or a particular field of a record can be used in an expression. An example of a record type declaration and of its use is shown in Figure 4-6.

```
TYPE atm_cell_header IS RECORD
    gfc : std_logic_vector( 3 DOWNTO 0 );
    rf  : std_logic_vector( 23 DOWNTO 0 );
    pt  : std_logic_vector( 2 DOWNTO 0 );
    clp : std_logic;
    hec : std_logic_vector( 7 DOWNTO 0 );
END RECORD;
```

```
VARIABLE cell_in, cell_out, cell_pkt : atm_cell_header;
VARIABLE vpi : std_logic_vector( 7 DOWNTO 0 );
VARIABLE vci : std_logic_vector( 15 DOWNTO 0 );
```

```
tmp_cell := cell_in;
cell_out.clp := cell_in.clp OR priority;
vpi := cell_in.rf( 23 DOWNTO 16 );
vci := cell_in.rf( 15 DOWNTO 0 );
```

Figure 4-6: Record type declaration and use

Behavioral synthesis tools process record types by splitting them into individual elements. This is the equivalent of the user representing each element of a record as an individual signal or variable. Thus, the record type is merely a convenience to the designer: no additional logic is implied when using a record in the place of multiple signals or variables. If a field within a record is also a record, it likewise is split into individual fields.

Figure 4-7 shows how a record is processed by synthesis.

```
TYPE atm_cell_header IS RECORD
    cfg : std_logic_vector( 3 DOWNTO 0 );
    rf  : std_logic_vector( 23 DOWNTO 0 );
    pt  : std_logic_vector( 2 DOWNTO 0 );
    clp : std_logic;
    hec : std_logic_vector( 7 DOWNTO 0 );
END RECORD;

SIGNAL header : atm_cell_header;
```

```
SIGNAL header_cfg : std_logic_vector( 3 DOWNTO 0 );
SIGNAL header_rf  : std_logic_vector( 23 DOWNTO 0 );
SIGNAL header_pt  : std_logic_vector( 2 DOWNTO 0 );
SIGNAL header_clp : std_logic;
SIGNAL header_hec : std_logic_vector( 7 DOWNTO 0 );
```

Figure 4-7: Record "splitting" during synthesis

> No additional hardware is generated when records are used in place of separate assignments, so records can be used freely in behavioral descriptions.

4.8 Array Type

One type of array has already been mentioned: the single-dimensional array. The types *bit_vector*, *std_logic_vector*, *signed*, and *unsigned* are all examples of single-dimensional arrays. Single-dimensional arrays are common in RTL descriptions and are used to store vectors of bits.

The VHDL language supports the specification of multi-dimensional arrays. This type of array can also be used in behavioral descriptions that will be synthesized.

Data Types

Behavioral synthesis tools can map 2-dimensional arrays to memory. Whether or not an array is mapped to a memory is under the direct control of the designer. By default, arrays are not mapped to memory. This section discusses the synthesis of arrays that are not being mapped to memory. Inferring memory in behavioral descriptions is discussed in Chapter 9.

Like records, arrays are used to group data together. Consider again the header of an Asynchronous Transfer Mode (ATM) cell. This field consists of 5 octets (bytes) of data. An array type could be defined to represent this data while it is being read off a data bus. The bytes could then be divided into the individual header fields in a record.

An array type declaration defines the number and type of elements in the array. Either an entire array or a particular element of an array can be used in an expression. An example of a 2-dimensional array type declaration and of its use is shown in Figure 4-8.

```
SUBTYPE octet IS std_logic_vector( 7 DOWNTO 0 );
TYPE atm_header_cell_data IS ARRAY ( 4 DOWNTO 0 ) OF octet;
```

```
VARIABLE header_data : atm_cell_header_data;
VARIABLE header      : atm_cell_header;
```

```
header.gfc := header_data( 4 )( 7 DOWNTO 4 );
header.rf  := header_data( 4 )( 3 DOWNTO 0 ) &
              header_data( 3 ) &
              header_data( 2 ) &
              header_data( 1 )( 7 DOWNTO 4 );
header.pt  := header_data( 1 )( 3 DOWNTO 0 );
header.clp := header_data( 1 )( 0 );
header.hec := header_data( 0 );
```

Figure 4-8: Array type declaration and use

Just like records, behavioral synthesis tools process array types (that are not mapped to memory) by splitting them into individual elements. This is equivalent to representing each element of an array as an individual signal or variable. Figure 4-9 shows how an array type is processed by synthesis.

Figure 4-9 shows only constant indexing of the array. When the array index is a constant, the synthesis tool can simply replace the index expression with the name of the signal that was generated during the "splitting" of the array. For example: **header_data(0)** is replaced with **header_data_0**.

> *No additional hardware is generated when multi-dimensional arrays are assigned or referenced with constant index expressions.*

```
SUBTYPE octet IS std_logic_vector( 7 DOWNTO 0 );
TYPE atm_header_cell_data IS ARRAY ( 4 DOWNTO 0 ) OF octet;

SIGNAL header_data : atm_cell_header_data;
```

```
SIGNAL header_data_4 : octet;
SIGNAL header_data_3 : octet;
SIGNAL header_data_2 : octet;
SIGNAL header_data_1 : octet;
SIGNAL header_data_0 : octet;
```

Figure 4-9: Array "splitting" during synthesis

When the array index is not a constant, additional hardware must be generated.

A non-constant *reference* to an element in an array implies a multiplexer. A reference is generated when non-constant indexing appears in an expression, such as on the right-hand side of an assignment statement. Depending upon the value of the index expression, any element in the array can be selected. Thus, the inputs to the multiplexer are every element in the array and the select line of the multiplexer is the non-constant index. An example of a non-constant index reference into an array and the hardware that will be generated by synthesis is shown in Figure 4-10.

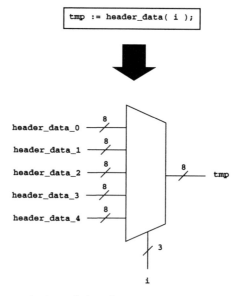

Figure 4-10: Non-constant array index reference and hardware generated by synthesis

Note that the 8-bit, 5-to-1 multiplexer required to implement the non-constant indexing of this small array represents a significant amount of hardware.

A non-constant *assignment* to an element in an array implies a decoder and a series of multiplexers. Such an assignment implies that the value of each element in the array may or may not be modified, depending upon the value of the index expression.

To represent this functionality in hardware, a multiplexer is required for each element of the array. The inputs to each multiplexer will be the "current" value of the array element and the possible new value. The select line of each multiplexer will be the output of a decoder that indicates which element is being modified. An example of a non-constant index assignment to an array and the hardware that will be generated by synthesis is shown in Figure 4-11. Note that the five 8-bit 2-to-1 multiplexers and the 3-bit decoder required to implement the non-constant indexing of this small array also represent a significant amount of hardware.

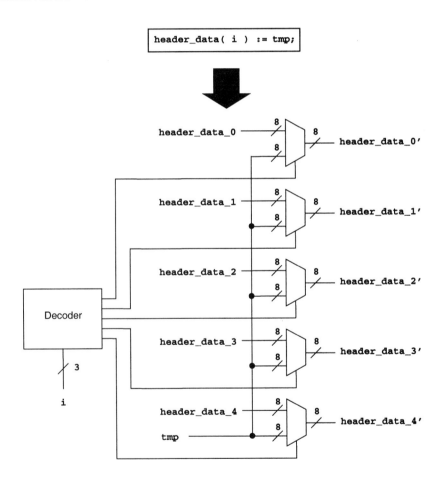

Figure 4-11: Non-constant array index assignment and hardware generated by synthesis

A synthesis tool may generate a more compact logic than that shown, but some form of hardware that performs this functionality must be constructed.

With non-constant indexing (both references and assignments), a very small amount of VHDL code can imply a vast amount of hardware, depending upon the size of the array being manipulated. The use of multi-dimensional arrays in behavioral descriptions can inadvertently result in large implementations.

> *Large amounts of hardware can be generated when a multi-dimensional array is assigned or referenced with a non-constant index expression. Such use of arrays should be carefully evaluated.*

The designer must consider the hardware that will be generated when using an array in a design. Large arrays whose elements are referenced or assigned with non-constant values should almost always be mapped to memory.

4.9 Types Not Supported for Synthesis

Certain types are not supported for behavioral synthesis. These types are also unsupported for RTL synthesis.

- Real Type

    There are many possible representations of floating point numbers. Today, any design that requires floating point numbers must have the representation of those numbers explicitly coded.

- Access Type (pointer)

    The access type is the equivalent to the pointer object found in most programming languages. Pointers are used to reference dynamically-generated objects. Today's synthesis tools do not support the dynamic creation of objects. It may be possible to support access types in the future as more hardware-software co-design tools are developed.

- Physical Type (e.g. time)

    Manipulation of physical types does not have a hardware equivalent.

    Certain uses of physical types (such as time specifications in AFTER clauses) are ignored by synthesis.

4.10 Summary

The VHDL language supports abstract types and non-abstract types, both of which can be used in a behavioral description. Non-abstract types are those types that easily represent wires and do not have to be modified by synthesis. Abstract types must be modified during synthesis in order to represent wires.

Although it may seem reasonable to use only non-abstract types when describing hardware, there are advantages to using abstract types in behavioral descriptions.

The use of abstract data types can simplify the coding of a design and its future maintenance. Abstract data types can simplify analysis and debugging. And abstract types can provide greater flexibility during synthesis.

Abstract types represent a powerful feature of the VHDL language. They are certainly appropriate for use in designs that are written at higher levels of abstraction. But as with any construct that is supported by synthesis, it is important to understand the effect of abstract types on synthesis and the overall design process.

Chapter 5

Entities, Architectures, and Processes

5.1 The Entity Declaration

In VHDL, the interface for a design is specified by the ENTITY declaration.

> *The ENTITY declaration specifies the input, output, and bi-directional signals that the design uses to communicate with external models. The ENTITY declaration also specifies any GENERIC values that can be used to parameterize the model.*

VHDL is a strongly typed language. Every port declared in an entity has an associated mode (direction) and type. A port can have any of the types that are supported for behavioral synthesis, as described in Chapter 4.

Like RTL synthesis, behavioral synthesis supports ports of mode IN, OUT, INOUT, and BUFFER. Ports of mode IN and OUT are input and output ports, respectively. Ports of mode INOUT are bi-directional ports: these ports can be driven from either inside or outside the design. Ports of mode BUFFER are also output ports, but the value of the port can be read inside the design. The VHDL language defines a port of mode OUT in a somewhat restrictive manner. A port of mode OUT is a signal that can only have values assigned: the value can not be read. Thus, the following statement is illegal if the port **dataout** is of mode OUT.

```
dataout <= dataout + 1;
```

This is illegal because the value of **dataout** is "read" on the right-hand side of the assignment. Ports of mode OUT can not be read. Changing the mode to BUFFER makes this statement legal and provides the desired functionality.

5.1.1 Ports With Abstract Types

A behavioral description usually represents an entire ASIC or a module within an ASIC. In either case, the design will need to interface to some other model, be it a test bench or another module in the design.

If an interface contains ports with abstract types (such as an enumeration type or an integer), those ports will be changed to non-abstract types during synthesis. Recall that non-abstract types are those types that easily represent wires and do not have to be modified by synthesis. In order to represent wires in the resultant design, abstract types must be transformed into non-abstract types by synthesis. The table in Figure 5-1 summarizes how the data types of signals and variables are modified by synthesis.

Pre-Synthesis Type	Post-Synthesis Type	Interface Modified
std_logic	std_logic	No
std_logic_vector	std_logic_vector	No
signed	signed	No
unsigned	unsigned	No
bit	std_logic	Yes
bit_vector	std_logic_vector	Yes
boolean	std_logic	Yes
integer	signed / unsigned	Yes
enumeration	std_logic_vector	Yes
RECORD	(record fields are "split")	Yes
2-D ARRAY	(array elements are "split")	Yes

Figure 5-1: Modification of data types during synthesis

If the interface to a design is modified by synthesis, the resultant netlist can no longer communicate with the surrounding environment, such as the test bench or another behavioral module in the design.

Consider a design that has an output port of type *boolean*, which is an abstract type. The type of this port will be changed to *std_logic* by synthesis. Without modification, a test bench that instantiated the original behavioral design can not be used to test the synthesized design: the type of the output port in the synthesized design (*std_logic*) can not be connected to the signal in the test bench (*boolean*). Such a connection is illegal in VHDL. This problem is illustrated in Figure 5-2.

This issue can be addressed in a number of ways.

One solution is to maintain two versions of the test bench -- one for the behavioral description and one for the synthesized description. Clearly, this approach is error-prone and creates extra work. If a change is made to one test bench, the same change would be needed in the other test bench. It would be easy for the two models to become unsynchronized.

Entities, Architectures, and Processes

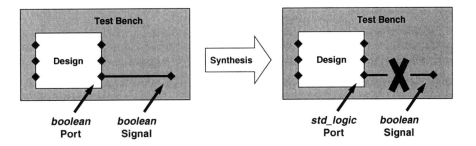

Figure 5-2: Type mismatch between design and test bench after synthesis

Alternately, a designer could avoid this issue entirely by only using the types *std_logic, std_logic_vector, signed,* and *unsigned* on ports. This would restrict the types that could be used in an interface to those that are not modified by synthesis. Inside the behavioral description, the designer could declare signals of the corresponding abstract types and perform an immediate conversion. An example of this is shown in Figure 5-3.

```
                                LIBRARY ieee;
                                USE ieee.std_logic_1164.ALL;
                                USE ieee.std_logic_arith.ALL;

                                ENTITY test IS
                                    PORT( clk : IN  std_logic;
                                          rst : IN  std_logic;
Describe interface using              a   : IN  signed( 3 DOWNTO 0 );
only non-abstract types               b   : IN  signed( 3 DOWNTO 0 );
                                          z   : OUT signed( 4 DOWNTO 0 ) );
                                END test;

                                ARCHITECTURE behavioral OF test IS

Declare internal signals            SIGNAL a_int : integer RANGE -8 TO 7;
with desired abstract               SIGNAL b_int : integer RANGE -8 TO 7;
types                               SIGNAL z_int : integer RANGE -16 TO 15;

                                BEGIN
Call conversion functions
in concurrent signal                a_int <= conv_signed( a, 4 );
assignments to map                  b_int <= conv_signed( b, 4 );
abstract to non-abstract            z <= conv_integer( z_int );
types and visa-versa
                                    -- Use a_int, b_int, and z_int for
                                    -- internal calculations
```

Figure 5-3: Design style that uses non-abstract interface types and abstract internal types

> *Avoid issues associated with type mismatches by only using types that are not changed by synthesis on ports. These types are std_logic, std_logic_vector, signed, and unsigned.*

5.1.2 Wrapper Models

Another possible solution to this problem involves the creation of a "wrapper" model that is placed between the test bench and the synthesized design. The sole purpose of the wrapper model is to preserve the original interface of the behavioral model and convert the data on these ports in the same manner as the synthesis tool.

The wrapper model has the same interface as the original design so that it can be directly instantiated in the test bench. This model instantiates the synthesized design. The model must declare internal signals with types that can be connected to the synthesized model. Conversion functions are used in the wrapper model to convert the types of the wrapper model to the types that appear in the synthesized design.

Figure 5-4 depicts the relationship between a test bench, a design that has abstract ports, and a wrapper model. The figure shows the relationships both before and after synthesis.

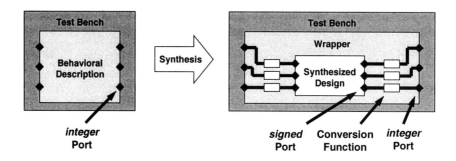

Figure 5-4: Modification of abstract design interface during synthesis

This solution does not prevent a designer from using abstract ports on the interface, but allows any synthesized designs to continue to communicate with other parts of the system.

> *If any of the ports of a design have data types that are modified during synthesis, a "wrapper" model can be created to allow the resultant design to communicate with the surrounding behavioral environment.*

Some synthesis tools can automatically generate a wrapper model when an input description has an interface with abstract types. This is certainly preferable to having to construct a wrapper model by hand. The generation of a wrapper model should be an easy job for the synthesis tool, since the tool is performing the transformations that result in the modification of the interface.

Hand generation of a wrapper model can introduce errors if the designer does not exactly replicate the transformations that are performed by synthesis. For many data types, it is simple to call the appropriate conversion function to connect the interface of the wrapper model to the interface of the synthesized design. However, an enumerated type is more difficult. In this case, the conversion function required for the wrapper model is specific to both the definition of the enumerated type and to the encoding method applied during synthesis (see Section 4.6 for more detail about how enumerated types are modified during synthesis).

Figure 5-5 shows an enumerated type, two possible encoding schemes, and the associated conversion functions. For enumerated types, any time there is a change to the type or the encoding scheme, the wrapper would need to be modified. To avoid these maintenance issues, the synthesis tool should generate the wrapper model.

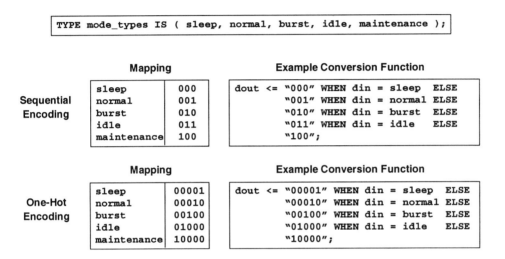

Figure 5-5: Encoding schemes for an enumerated type and associated conversion functions

To correctly connect the wrapper model and the synthesized design, the following conditions must be met:

- The name of the wrapper model is the name of the original design.
- The interface of the wrapper model is identical to the interface of the original design.
- The wrapper model instantiates the synthesized design.
- The name of the synthesized design is modified (so that it can be instantiated inside the wrapper model).
- The wrapper model calls appropriate conversion functions to translate data from the interface of the wrapper model to the interface of the synthesized design, and vice versa.

Figure 5-6 shows the interface of a simple design with abstract ports and the modified interface after synthesis. Figure 5-7 shows the wrapper that is required to simulate the synthesized design with the original test bench.

```
ENTITY test IS
  PORT(clk : IN  std_logic;
       rst : IN  std_logic;
       a   : IN  integer RANGE -8 TO 7;
       b   : IN  integer RANGE -8 TO 7;
       z   : OUT integer RANGE -16 TO 15 );
END test;
```

```
ENTITY test_syn IS
  PORT(clk : IN  std_logic;
       rst : IN  std_logic;
       a   : IN  signed( 3 DOWNTO 0 );
       b   : IN  signed( 3 DOWNTO 0 );
       z   : OUT signed( 4 DOWNTO 0 ) );
END test_syn;
```

Figure 5-6: Interface specification of a simple design before and after synthesis

Entities, Architectures, and Processes

```
LIBRARY ieee;
USE ieee.std_logic_1164.ALL;
USE ieee.std_logic_arith.ALL;

ENTITY test IS
    PORT( clk : IN  std_logic;
          rst : IN  std_logic;
          a   : IN  integer RANGE -8 TO 7;
          b   : IN  integer RANGE -8 TO 7;
          z   : OUT integer RANGE -16 TO 15 );
END test;

ARCHITECTURE wrapper OF test IS

    COMPONENT test_syn
        PORT ( clk : IN  std_logic;
               rst : IN  std_logic;
               a   : IN  signed( 3 DOWNTO 0 );
               b   : IN  signed( 3 DOWNTO 0 );
               z   : OUT signed( 4 DOWNTO 0 ) );
    END COMPONENT;

    SIGNAL clk_design : std_logic;
    SIGNAL rst_design : std_logic;
    SIGNAL a_design   : signed( 3 DOWNTO 0 );
    SIGNAL b_design   : signed( 3 DOWNTO 0 );
    SIGNAL z_design   : signed( 4 DOWNTO 0 );

BEGIN

    clk_design <= clk;
    rst_design <= rst;
    a_design <= conv_signed( a, 4 );
    b_design <= conv_signed( b, 4 );
    z <= conv_integer( z_design );

    u1: test_syn
        PORT MAP ( clk => clk_design,
                   rst => rst_design,
                   a   => a_design,
                   b   => b_design,
                   z   => z_design );
END wrapper;
```

- Wrapper model has same name and interface of original design
- Declare internal signals to hook up synthesized design
- Call conversion functions to map abstract to non-abstract types and visa-versa
- Instantiate synthesized design (note name change)

Figure 5-7: Wrapper model for a simple design

5.2 The Architecture Specification

In VHDL, an *architecture* describes the functionality of a design. An entity can have more than one associated architecture, however the interface (which is defined by the entity) will be identical for each architecture.

An architecture can contain concurrent signal assignment statements, component instantiations, and processes.

> *The architecture describes an algorithm to be implemented. An algorithm can be defined in VHDL using a combination of processes, concurrent signal assignment statements, and component instantiations. A typical RTL design contains all of these; a behavioral description can contain only a single process.*

Behavioral descriptions can contain component instantiations. Behavioral synthesis tools process component instantiations in the same manner as RTL synthesis: the component is placed directly into the resultant design. In RTL descriptions, instantiated components tend to be technology cells or "pre-designed" portions of a design.

Because behavioral design involves working at a more abstract level than RTL design, it is better to move such instantiations outside of the behavioral description. Inserting technology cells into a behavioral description is somewhat contradictory to the technology-independent nature of behavioral design.

Behavioral code can be separated from other portions of the design by creating a level of hierarchy that instantiates a component (that contains the behavioral code) along with other appropriate technology or pre-designed logic. Insulating the behavioral description also encourages reuse of that code in designs that may target a different technology, have different design goals, or require an alternate architecture.

> *Component instantiations are synthesized for behavioral synthesis in the same manner as RTL synthesis: the instances are passed through from the source design to the synthesized netlist.*

Behavioral descriptions can contain concurrent assignment statements. A concurrent assignment statement can represent an algorithm to be scheduled (as is the case in the example discussed in Chapter 2), or it can represent simple "glue" logic (such as gating logic or a simple type conversion).

Some behavioral synthesis tools view all concurrent assignment statements as representing "glue" logic and thus will not attempt to schedule the operations that appear in those statements. In this case, the statement is synthesized in the same manner as would be performed by an RTL synthesis tool.

Other behavioral synthesis tools translate each concurrent assignment statement into the equivalent process (every concurrent signal assignment can be represented by a functionally equivalent process) and then attempts to schedule the process. The

Entities, Architectures, and Processes

example discussed in Chapter 2 is a concurrent signal assignment that has been scheduled. Synthesis tools that do not schedule concurrent assignment statements by default provide a mechanism for the user to indicate that a particular statement should not be scheduled.

> *Concurrent signal assignment statements can be treated either as algorithms to be scheduled or as purely combinational logic.*

5.3 Processes

The role of a process in a behavioral description is very different than in an RTL description. In an RTL design, each process is either a "combinational" process or a "clocked" process. A combinational process will generate what its name implies: combinational logic. A clocked process will generate combinational logic and then place registers at each of the "outputs" of the process. Figure 5-8 shows an example of a combinational process and a clocked process.

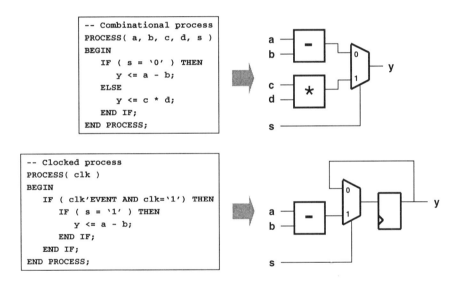

Figure 5-8: Combinational and clocked processes used in RTL design

When working at the RTL level, each process represents a building block. The designer must partition an algorithm into these blocks before any coding can begin. Because an RTL description is typically composed of many blocks, the overall structure of the algorithm is usually very difficult to see. All of the processes must be viewed together to understand the overall structure of the design.

Recall the algorithm discussed in Chapter 2:

$$y = a * b + c + (d - e) * f$$

Figure 5-9 depicts the structure of a possible RTL description for this algorithm. Each of the shaded areas could be coded as a separate combinational or clocked process.

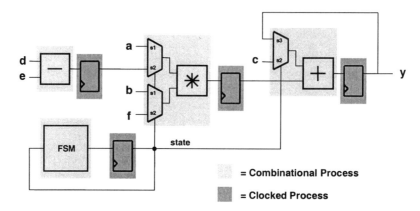

Figure 5-9: The structure of an RTL description of a simple algorithm

Figure 5-10 shows RTL-style VHDL code which could be used to implement the structure shown in Figure 5-9. By just examining this code it is quite difficult to see the simple equation which is being implemented.

```vhdl
ARCHITECTURE behavioral OF simple IS
   SIGNAL sub, sub_reg : signed( 31 DOWNTO 0 );
   SIGNAL mul, mul_reg : signed( 31 DOWNTO 0 );
   SIGNAL add, add_reg : signed( 31 DOWNTO 0 );

   TYPE state_type IS ( s1, s2, s3 );
   SIGNAL state, next_state : state_type;
BEGIN

   PROCESS( state )
   BEGIN
      CASE state IS
         WHEN s1 => next_state <= s2;
         WHEN s2 => next_state <= s3;
         WHEN s3 => next_state <= s1;
      END CASE;
   END PROCESS;

   PROCESS( clk )
   BEGIN
      IF ( clk'EVENT AND clk='1' ) THEN
         state <= next_state;
      END IF;
   END PROCESS;

   PROCESS( clk )
   BEGIN
      IF ( clk'EVENT AND clk='1' ) THEN
         sub_reg <= sub;
         mul_reg <= mul;
         add_reg <= add;
      END IF;
   END PROCESS;
```

```vhdl
   PROCESS( d, e )
   BEGIN
      sub <= d - e;
   END PROCESS;

   PROCESS( a, b, f, sub_reg, state )
      VARIABLE mul1 : signed( 31 DOWNTO 0 );
      VARIABLE mul2 : signed( 31 DOWNTO 0 );
   BEGIN
      IF ( state = s1 ) THEN
         mul1 := a;
         mul2 := b;
      ELSE
         mul1 := sub_reg;
         mul2 := f;
      END IF;
      mul <= mul1 * mul2;
   END PROCESS;

   PROCESS( a, b, f, y_reg, state )
      VARIABLE add1 : signed( 31 DOWNTO 0 );
   BEGIN
      IF ( state = s2 ) THEN
         add1 := c;
      ELSE
         add1 := y_reg;
      END IF;
      add <= add1 + mul_reg;
   END PROCESS;

   y <= add_reg;

END behavioral;
```

Figure 5-10: Portion of RTL-style VHDL code to implement the simple algorithm

Entities, Architectures, and Processes

Even though RTL descriptions are thought of as "technology independent", such descriptions are based on the characteristics of a particular ASIC or FPGA technology.

The path highlighted in Figure 5-11 implies that the multiply operation (along with a multiplexer operation) can be performed in a single clock cycle. Presumably, the designer thought about this when coding the design. If the designer did not consider the delay of a multiplier in a particular technology, he will be surprised when gate-level optimization is unable to meet the required timing constraints.

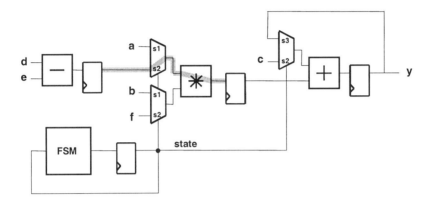

Figure 5-11: Implementation of a simple algorithm showing path through multiplier

> *The tasks that are performed by behavioral synthesis tools are not unknown to the RTL designer. An RTL design is a scheduled design. In the RTL design process, the designer must have already scheduled the design before writing any code, even though this may not be a very conscious undertaking. Thus, RTL design activities are architecture-dependent.*

Even if such a design can be optimized for one ASIC technology, it may not function in a slower technology (such as an FPGA technology). If the multiplier requires more than the delay of one clock cycle to perform the operation, the RTL code has to be re-written for the slower technology. An RTL description implies a single implementation architecture that makes assumptions about the target technology. If the code is re-written, the new RTL code describes a different hardware architecture which also makes assumptions about the target technology.

> *Even through RTL designs have the appearance of technology-independence, they are very technology-dependent. The issues associated with the technology (e.g. area and timing) are incorporated into the description by the designer at the time the RTL code is written.*

Unlike an RTL description, a behavioral description is truly technology-independent. The timing and area characteristics of components (such as multipliers) that will be used to implement the design need not be known when the behavioral code is written.

A behavioral description is also architecture-independent. Even the *number* of data path components that will be used to implement the design need not be known when the behavioral code is written.

An RTL description usually contains a number of processes, each one representing a part of the design. In a behavioral description, a process is used to represent the overall flow of an algorithm. For this reason, a behavioral description might contain only a single process to be scheduled.

Minimally, a process to be scheduled describes *data dependencies.* Consider again the simple equation that was discussed in Chapter 2. Recall how this equation implied a Data Flow Graph (DFG), as shown in Figure 5-12. A DFG shows the dependencies between data.

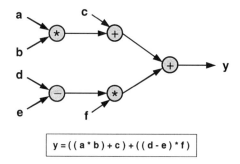

Figure 5-12: Simple equation and corresponding data flow graph (DFG)

The DFG shows the order in which operations must be performed. For example, the operation which multiplies the signals **a** and **b** must be completed before the result is added to **c**. This may seem obvious, but understanding data dependencies is essential to understanding behavioral synthesis. Chapter 2 discussed a number of different ways to schedule this equation based on varying constraints (such as clock period and available resources), but note that each solution adheres to the data dependencies described in the source code.

> *Data dependencies cannot be violated during behavioral synthesis.*

The flow of sequential statements from the start of a process to the end of a process corresponds to a flow of data from inputs to outputs in a DFG. Behavioral synthesis takes the operations that are described in a process and schedules those operations into multiple clock cycles. Since behavioral synthesis tools register outputs, a scheduled process is similar to a clocked process in RTL design. But there is a fundamental

Entities, Architectures, and Processes

difference between the two: scheduled processes can take many clock cycles to perform a calculation, an RTL clocked process can take only one.

> When using RTL synthesis, the operations in a process must complete in one clock cycle. When using behavioral synthesis, the operations in a process can take any number of clock cycles, depending upon user-specified constraints.

5.3.1 Specifying Clock Edges

In an RTL description, a clock edge can be specified using either an IF statement or a WAIT statement. The two processes in Figure 5-13 are functionally equivalent, in that in

```
-- Clocked process
PROCESS( clk )
BEGIN
    IF ( clk'EVENT AND clk='1') THEN
        IF ( s = '1' ) THEN
            y <= a - b;
        END IF;
    END IF;
END PROCESS;
```

```
-- Clocked process
PROCESS
BEGIN
    WAIT UNTIL clk'EVENT AND clk='1';
    IF ( s = '1' ) THEN
        y <= a - b;
    END IF;
END PROCESS;
```

Figure 5-13: Specification of a clock edge using an IF statement and a WAIT statement

both processes the value of the signal **y** is modified on each rising edge of the signal **clk**. The WAIT statement in the example represents waiting for a rising edge. Replacing **clk='1'** with **clk='0'** specifies waiting for a falling edge.

Clock edges can also be specified using the functions *rising_edge* and *falling_edge* as defined in the *std_logic_1164* package. Unlike most other clock edge definitions, these functions check the last value of the clock prior to the transition to ensure that clock edge transitions start at '0' or '1' (for rising and falling edges, respectively). Transitions that begin at other values (such as 'X' or '-') are not valid clock edges. An example of a WAIT statement that uses the *rising_edge* function is:

```
WAIT UNTIL rising_edge( clk );
```

Clock edge specifications that use the *rising_edge* and *falling_edge* functions are better simulation models.

During simulation, when a process contains only a single WAIT statement, the code that appears in a process is executed in a single clock cycle. Similarly, a process that contains multiple WAIT statements will require multiple clock cycles to execute. Each WAIT statement in a process can be thought of as a state in a state machine, as shown in Figure 5-14.

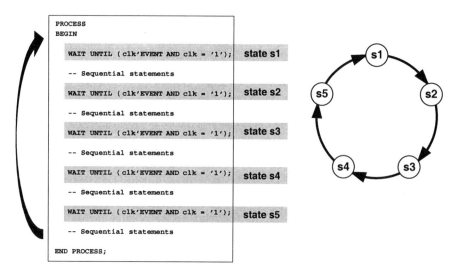

Figure 5-14: Correlation between multiple WAIT statements and a state machine

Behavioral synthesis tools allow processes to contain multiple WAIT statements. The WAIT statements must all represent rising or falling edges of the same clock signal. Behavioral synthesis does not allow a process to contain a mix of rising and falling clock edges.

> In a behavioral description, a process is used to describe an entire (or large portion of an) algorithm. A process can contain multiple WAIT statements that imply a state machine. The state machine is automatically generated during synthesis.

5.3.2 Timing Diagrams

WAIT statements are used in behavioral models to correlate the operation of an algorithm to a clock. In simulation models, the WAIT statements indicate when input signals are read and when output signals are written. The total number of WAIT statements in the process indicates how long it takes to calculate the algorithm.

The most common representation of the correlation between an algorithm and a clock is a timing diagram. A timing diagram is a graphical representation of the I/O timing requirements of the design. A behavioral description can describe more than just functionality -- it can also describe timing information. One goal of a behavioral description is to allow accurate simulation of the I/O behavior.

I/O timing requirements might stem from two sources:

1. The surrounding design modules dictate when certain signals are available, or should become available, to the design under creation.

Entities, Architectures, and Processes

2. The timing is described in the functional specification, with the external world or other design modules in mind.

In either case the designer is often not allowed to modify these requirements.

Consider a timing diagram shown in Figure 5-15 for the algorithm discussed in Chapter 2.

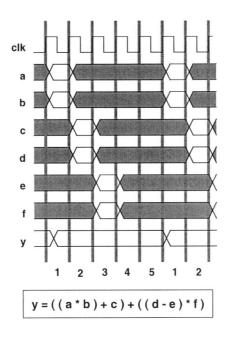

Figure 5-15: I/O timing diagram for a simple algorithm

The timing diagram indicates that the inputs **a** and **b** become valid shortly after the rising edge of clock cycle 1 and are held stable until just past the end of that cycle. This implies that these signals must be sampled during the first clock cycle. Similarly, the inputs **c** and **d** are valid during the second clock cycle and **e** and **f** are valid during the third clock cycle.

The output signal, **y**, becomes valid shortly after the clock edge which ends the fifth clock cycle but is held valid until its value changes again. This implies that the implementation of this algorithm will have a *registered output* signal.

The timing diagram indicates that the entire algorithm is calculated in 5 clock cycles, at which point a new calculation begins.

> *The latency of an algorithm is the number of clock cycles required to perform the overall calculation.*

Understanding Behavioral Synthesis

This algorithm and the I/O timing shown in Figure 5-15 can be modeled in a single process with the use of multiple WAIT statements. The model will describe both the algorithm and the I/O timing for the inputs and outputs.

The algorithm that was discussed in Chapter 2 was modeled using a concurrent signal assignment. This VHDL model did not take any I/O timing into consideration. A new output value was calculated immediately whenever an input value changed. The scheduling examples that were examined showed how actual hardware that implements the algorithm would take one or more clock cycles to complete.

> *It is better to write VHDL code that takes the I/O timing specification into account. This allows the algorithm to be accurately verified in the context of the surrounding system while it is being developed, prior to synthesis.*

Figure 5-16 shows VHDL code that represents the I/O timing diagram.

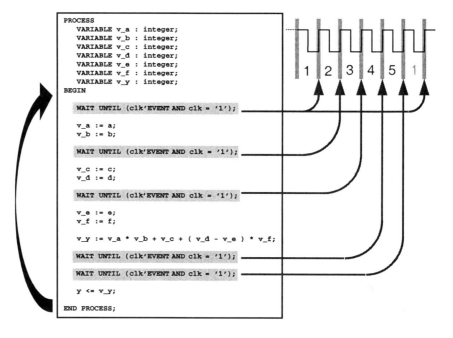

Figure 5-16: VHDL code that implements the I/O timing diagram for the algorithm

The algorithm is described within a single process. Clock cycles are separated in the code using WAIT statements. Each WAIT statement describes the rising edge of the signal **clk**.

Entities, Architectures, and Processes 73

> *WAIT statements can be thought of as representing states in a state machine. The assignments between the WAIT statements represent the actions that occur when transitioning from one state to the next.*

The process begins by waiting for the rising edge of the clock and then assigns the values of the input signals **a** and **b** to corresponding variables, **v_a** and **v_b**. The values are stored in these variables for later use since the timing diagram indicates that the signals are only valid during the first clock cycle.

The statements that *follow* a WAIT statement describe the events that should happen at that clock edge during the transition from one state to the next. For example, the signals **a** and **b** are sampled at the clock edge that ends cycle 1.

Similar to the signals **a** and **b**, the signals **c** and **d** are also stored in variables for later. The timing diagram indicates that these signals are only valid during the second clock cycle. The values of the signals **e** and **f** are sampled at the end of the third clock cycle.

Once all of the input values have been sampled, the algorithm can be calculated. The actual assignment to the output signal **y** follows a WAIT statement that corresponds to cycle 5. The timing diagram indicates that the value of **y** should change on the clock edge between cycle 5 and cycle 1 (which is the start of the next calculation).

When the end of the process is reached, simulation continues at the start of the process.

5.3.3 Scheduling Assumptions

The timing that is shown in the diagram is characteristic of a synchronous system. The input signals are shown as changing shortly after a clock edge. In hardware, this implies that these signals have been registered outside of the model being designed. The short delay between the clock edge and the change to the input signal may correspond to the clock-to-output delay of an external register. The output signal is shown as changing shortly after a clock edge. Similarly, this implies that this signal will be registered in the hardware implementation of the algorithm.

Behavioral synthesis tools assume that a process that is being scheduled is part of a synchronous system. These assumptions about input signals and output signals are illustrated in Figure 5-17.

> *Behavioral synthesis makes assumptions about the I/O timing of any processes being scheduled. Synthesis assumes that the inputs to a process are externally registered and will thus be stable for an entire clock cycle. Similarly, synthesis will register the outputs of a scheduled process; output values will only change at clock boundaries.*

A behavioral description usually contains a single process that reads from and writes to interface signals (ports). But a behavioral description can contain two or more processes. When a description contains multiple processes, data is passed between processes using internal signals.

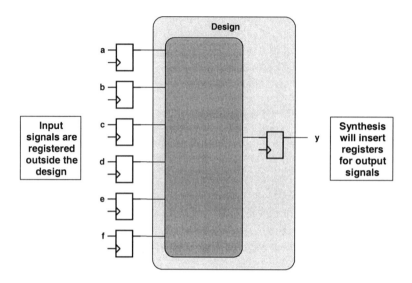

Figure 5-17: I/O assumptions for synchronous systems

Variables can be used within a process. But variables are local to a process and can not be read from or written to outside of the process in which they are declared.

The structure of a design that contains multiple processes is shown in Figure 5-18. Processes read from and write to signals. The variable **v1** in process **p1** and the variable **v1** in process **p2** are unrelated.

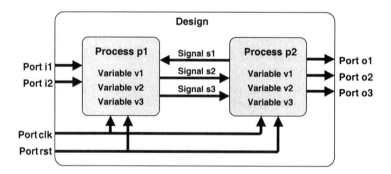

Figure 5-18: General structure of a design with multiple processes

Behavioral synthesis tools schedule every process in a design separately. The I/O assumptions that were previously discusses are applied to the "interface" for every process in the design. The interface of a process is composed of the set of signals that are read from or written to in the process. These signals can be interface signals (ports) or they can be internal to the design, as shown in the figure.

Entities, Architectures, and Processes

> *Because each process is scheduled independently, reading from or assigning to an internal signal is treated by behavioral synthesis in the same manner as an interface signal (port). This means that all internal signals are registered and that data passes between processes only on clock edges.*

If an internal signal is written to in a process, it will be registered in the same manner as an output port. This scenario is illustrated in Figure 5-19.

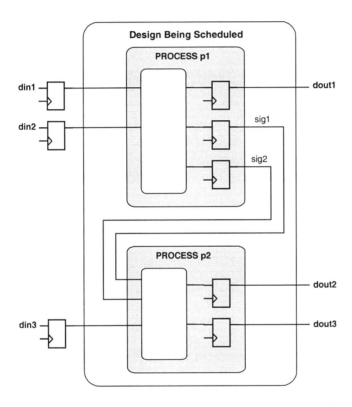

Figure 5-19: Illustration of scheduling assumptions applied to a design with multiple processes

5.4 Summary

Entities, architectures, and processes represent the fundamental building blocks of any VHDL design. It is important to understand how a behavioral synthesis tool will process each of these constructs.

Synthesis modifies data types. The use of particular data types in an entity declaration can have significant impact on the design flow. In particular, the use of abstract types on an interface will result in the modification of that interface during synthesis. It is necessary to design a "wrapper" model to perform simulation of a design that was produced by synthesis in the context of a test bench or a larger system.

A VHDL architecture describes the functionality of the design. In RTL design, an architecture may contain many concurrent assignment statements, component instantiations, and processes. In behavioral design an architecture may contain only a single process.

While entities and architectures are treated by behavioral synthesis in a manner that is very similar to RTL synthesis, the role of a process is very different. In RTL design, the design is segmented by the designer -- many processes are used to describe the design. In behavioral design, a process is used to describe an entire algorithm.

A process can include multiple clock edges that represent the I/O timing of the design. These clock edges imply a state machine that is automatically constructed during synthesis.

Chapter 6

Loops

6.1 Loops in RTL Design

To help understand how loops are used in behavioral descriptions, it is useful to review how loops can be used in RTL descriptions. Many loop structures imply a state machine that is required to control the operation of the loop. For example, when the number of times a loop executes depends on the input data, some kind of a state machine is required. Historically, RTL synthesis tools have not automatically created state machines.

Loops do not appear very often in RTL descriptions. This is largely because loops have generally only supported in a very limited manner for RTL synthesis – usually only FOR loops with constant bounds can be specified. FOR loops with non-constant bounds, WHILE loops, and infinite loops have not been supported or supported only in a limited manner for RTL synthesis.

Many RTL designers are used to writing RTL code based on this limitation. Some examples of loops that illustrate these restrictions are shown in Figure 7-1.

```
FOR i IN 0 TO 7 LOOP
    -- Loop contents
END LOOP;

FOR i IN 15 DOWNTO 7 LOOP
    -- Loop contents
END LOOP;

FOR v IN -10 TO 10 LOOP
    -- Loop contents
END LOOP;
```

(a) Legal for RTL synthesis

```
FOR i IN 1 TO k LOOP
    -- Loop contents
END LOOP;

WHILE (done = '0') LOOP
    -- Loop contents
END LOOP;

LOOP
    -- Loop contents
END LOOP;
```

(b) Not legal for RTL synthesis

Figure 7-1: Examples of loops

During RTL synthesis, FOR loops are *unrolled*. To unroll a FOR loop, the statements inside the loop are copied as many times as the loop would be executed. For each copy of the loop statements, the iteration variable is replaced with the appropriate constant for that iteration of the loop.

Figure 7-2 shows a FOR loop with constant bounds and the equivalent code after the loop has been unrolled.

```
FOR i IN 0 TO 6 LOOP
    s( i ) := s( i + 1 );
END LOOP;
```

Unrolling

```
s( 0 ) := s( 1 );
s( 1 ) := s( 2 );
s( 2 ) := s( 3 );
s( 3 ) := s( 4 );
s( 4 ) := s( 5 );
s( 5 ) := s( 6 );
s( 6 ) := s( 7 );
```

Figure 7-2: Unrolling a FOR loop with fixed bounds

This transformation implies that in an RTL synthesis tool that doesn't perform resource sharing, an unrolled FOR loop always results in the generation of all of the hardware that is required to perform the execution of *every* iteration of the loop in the resulting design. A few lines of code can generate significant amounts of hardware. For example, if a FOR loop in an RTL description that is executed 100 times contains a statement with a 16-bit add operation, the circuit generated by synthesis will contain 100 adder components! If an RTL designer wishes to use a single adder and *schedule* the calculation over 100 clock cycles, he would have to write code that describes a state machine and the appropriate surrounding circuitry to multiplex data through the adder and to store intermediate values.

Consider the FOR loop in Figure 7-3. When unrolled, this loop not only implies 7 adders, it also introduces significant delay into the synthesized design.

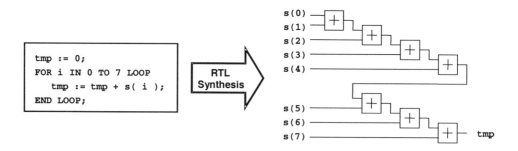

Figure 7-3: Hardware inferred from a FOR loop with fixed bounds

Loops

> *Most RTL synthesis methodologies discourage the use of FOR loops. Because many RTL synthesis tools always unroll FOR loops, it is easy to generate a significant amount of unnecessary hardware with a loop.*

The main use of loops in RTL design is to provide a convenient way to code simple repetitive structures as shown in Figure 7-2.

In addition to limiting the use of loops to FOR loops with fixed bounds, most RTL synthesis tools do not support NEXT and EXIT statements. For these reasons, loops rarely appear in RTL descriptions.

6.2 Loop Constructs and State Machines

In order to support more complex looping structures, a synthesis tool must be able to generate state machines. RTL tools have not traditionally done this. Some of today's RTL synthesis tools support the use of multiple WAIT statements, but only in a very simple way.

Chapter 5 discussed how a process implies a state machine – WAIT statements can be thought of as corresponding to states and the statements between the WAIT statements can be thought of as occurring on the transition between states.

Figure 7-4 illustrates the relationship between WAIT statements in a PROCESS and an implied state machine.

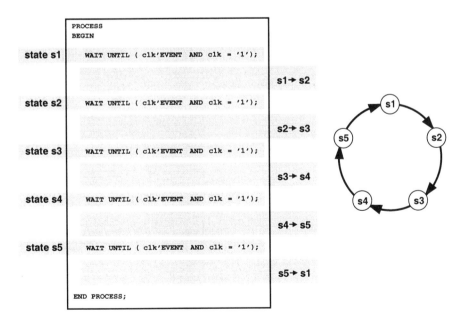

Figure 7-4: Relationship between WAIT statements and a state machine in a process

In VHDL simulation, when the last statement in a process has been executed, simulation continues at the first statement in the process. In the state machine that corresponds to this description, there is a transition from the last state (**s5**) back to the first state (**s1**). A process infers a "circular" state machine. This means that any design that is scheduled will perform a series of operations over and over again.

An *infinite loop* also implies a "circular" state machine. Figure 7-5 shows VHDL code that is almost identical to the code in Figure 7-4. However, the repetitive nature of this state machine is inferred by a *loop* construct, rather than a process.

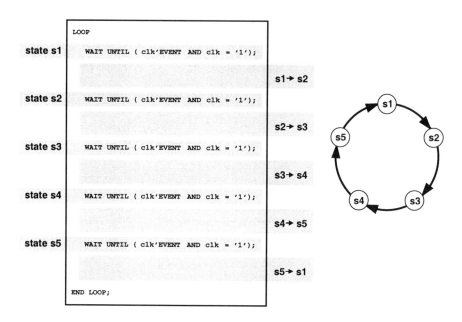

Figure 7-5: Relationship between WAIT statements and a state machine in an infinite loop

This loop is called an *infinite loop*, because the loop construct itself does not describe any conditions for exiting the loop. An infinite loop implies a circular state machine. Branches can be added to a state machine using NEXT statements (discussed in Section 6.5) and EXIT statements (discussed in Section 6.3). In addition to infinite loops, there are FOR loops and WHILE loops, which describe conditions for exiting the loop.

The loop construct is fundamental to writing behavioral descriptions. Loops provide a powerful mechanism for describing repetitive operations that occur in an algorithm. A goal of behavioral synthesis is to allow designers to describe at a higher level of abstraction that is closer to the algorithm. This would be very difficult if the use of loop structures were disallowed since algorithms often use looping structures to describe their function.

Loops

6.3 The EXIT Statement

The EXIT statement transfers control to the statement that follows the end of a loop. Any statements between the EXIT statement and the end of the loop are not executed.

There are a number of different ways to use an EXIT statement. An exit from a loop can be conditional or unconditional. In addition, an EXIT statement can specify the name of the loop to exit -- this provides a mechanism for exiting an outer loop from within an inner loop. If a loop name is not specified, an EXIT statement will result in exiting the inner-most loop. All of these forms can be used in behavioral descriptions. Figure 7-6 shows examples of code fragments that include EXIT statements.

```
IF ( exit_condition = '1' ) THEN          EXIT WHEN exit_condition = true;
    EXIT;
END IF;

IF ( exit_condition = '1' ) THEN          EXIT outer_loop WHEN sum > 25;
    EXIT outer_loop;
END IF;
```

Figure 7-6: Example uses of the EXIT statement

Figure 7-7 shows two nested infinite loops with an EXIT statement that conditionally exits the inner loop. The figure also shows the state machine that corresponds to this code.

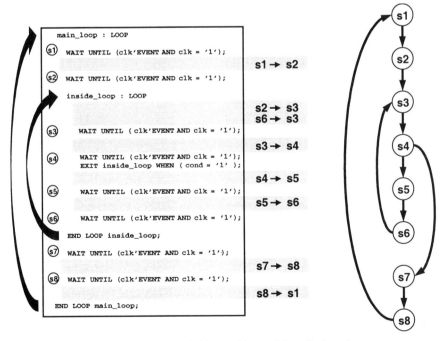

Figure 7-7: Two embedded loops with an exit from the inner loop

To understand this diagram, think of the WAIT statements as representing states and the code between WAIT statements as the transitions between states.

To construct the state diagram that this code represents, every possible state-to-state transition must be considered. For example, the statements that immediately follow the start of the loop called "internal_loop" are executed under two possible conditions:

> ... when transitioning from state **s2** to **s3** when entering the loop

> ... when transitioning from state **s6** to **s3** on successive iterations of the inner loop

The EXIT statement creates branches in the state machine. The transition from state **s4** to **s5** is taken when the condition that controls the EXIT statement is false. The transition from state **s4** to state **s7** is taken if that condition is true.

This implies that behavioral synthesis tools can construct Mealy-style state machines. In such state machines, the output assignments are a function of both the current state and other inputs. For example, the assignments that follow the EXIT statement are only performed if the state machine is in state **s4** and the value of the input signal **cond** is '0'.

> *Loops with EXIT statements can be used in behavioral descriptions to affect the state machine that is created during synthesis. An EXIT statement will correspond to a branch in the generated state machine.*

6.3.1 Specifying a Synchronous Reset

An important consideration when designing a circuit is providing a method for bringing the circuit into a known state. This is typically done using a reset signal that is made active for a short period of time when power is applied to the circuit. Since in most technologies the power-up state of registers is not predictable, resetting a circuit usually involves setting output and internal registers to either a '0' or a '1'.

Any scheduled design that implies a state machine *must* contain a reset signal. This is because the reset signal is used to initialize the state machine. The reset signal can also be used to initialize other portions of the circuit, but without a reset, the gate-level implementation of the state machine can not be set to a known state.

In an RTL description, a synchronous reset is described by testing a reset condition at the clock edge. A reset condition is a comparison between a reset signal and the value '0' or '1' (for an active-low or active-high reset, respectively). If the reset condition is true, then the assignments that comprise the *reset action* are performed. A reset action consists of assigning constant values (zeros and/or ones) to one or more signals or variables. If the reset condition is false, the behavior of the circuit continues unaffected.

The same behavior must be specified in a behavioral description: at the clock edge, the reset condition must be tested. The difference between an RTL clocked process and a behavioral process is that the behavioral process can describe multiple clock edges. So in a behavioral description, it is necessary to test the reset condition at *every* clock edge. A description with multiple WAIT statements implies a state machine with multiple states. A design with a synchronous reset must have a transition from each of these states back

Loops

to the first state. Since behavioral synthesis tools can create Mealy-style state machines, the reset actions can be thought of as occurring on the transitions from each state back to the first state. Such transitions can be created with EXIT statements.

Similar to an RTL clocked process, the reset action in a behavioral description must be separated from the normal operation of the circuit. After a circuit has been reset, normal operation should resume. The circuit should not perform the reset action again unless the reset condition is once again true.

> *To define a synchronous reset for a behavioral design, an EXIT statement must follow every WAIT statement. These EXIT statements must all have the same exit condition and should exit an infinite loop that describes the normal (non-reset) behavior of the design. The reset behavior is described outside the main loop.*

6.3.2 Simulating the Reset Condition

The VHDL code to describe a synchronous reset is shown in Figure 7-8.

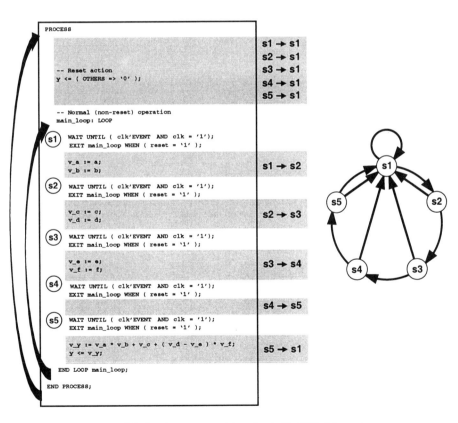

Figure 7-8: Synchronous reset described with EXIT statements

Consider how this description will simulate when the reset condition becomes true (i.e. the value of the signal **reset** is '1'). Assume that the active statement is the WAIT statement that corresponds to state **s3**. The next time there is a rising clock edge, execution will proceed to the EXIT statement which follows the WAIT statement. Since the reset condition is now true, execution will continue with the first statement that follows the loop labeled "main_loop". But there are no statements after the loop, so the process repeats from the top. This is where the reset action is defined.

In this example, the reset action assigns zeros to the output signal **y**. Execution continues to the next statement, which is the first WAIT statement inside the main loop. If the reset action is still true at the next rising clock edge, the main loop is exited once again. As long as the reset action is true, only the assignments in the reset action will be performed. The circuit is "stuck" in state **s1**. When the reset action becomes false, the main loop continues executing normally. Since the main loop is an infinite loop, when the last statement in the main loop is reached, execution continues at the top of that loop.

Now consider how the VHDL code in Figure 7-8 will simulate in general. Simulation begins at the first statement in the process, which is the reset action. The reset action is performed and the loop that describes the normal operation of the circuit is entered. This means that the reset action is *always* performed at the start of VHDL simulation – even if the reset condition was never set to true!

A better behavioral model would require that the reset condition be true in order for the reset action to be performed. This can be accomplished by moving the reset action from *before* the main loop to *after* the main loop. The two methods of specifying a synchronous reset are shown in Figure 7-9.

```
PROCESS

    -- Reset action before the main loop is
    -- performed even if reset condition is
    -- NEVER activated!

main_loop: LOOP
    -- Normal (non-reset) operation

    WAIT UNTIL (clk'EVENT AND clk = '1');
    EXIT main_loop WHEN ( reset = '1' );

    -- Algorithm goes here

END LOOP main_loop;

END PROCESS;
```

```
PROCESS

main_loop: LOOP
    -- Normal (non-reset) operation

    WAIT UNTIL (clk'EVENT AND clk = '1');
    EXIT main_loop WHEN ( reset = '1' );

    -- Algorithm goes here

END LOOP main_loop;

    -- Reset action after the main loop is
    -- ONLY performed if reset condition
    -- is activated!

END PROCESS;
```

Figure 7-9: Alternate positions for reset action – before (left) and after (right) the main loop

The structure on the right requires that the reset condition be true in order for the reset action to be performed. This means that a test bench must exercise the reset condition even when simulating the behavioral model. This is important since a synthesized design will contain a state machine that *must* be reset. Placing the reset action at the end of a behavioral model helps ensure that a test bench or any other model that is interfacing with the description will perform an appropriate reset.

Loops

> *It is best to place the reset action after the main loop (instead of before). Although this is a slightly less intuitive code structure (the reset action is the last thing in the behavioral description), the behavioral description will simulate more like the synthesized design.*

6.3.3 Complex Reset Sequences

A reset action can only contain simple assignments of constants – there can not be any expressions that require actual time to compute (such as an add). But some circuits will require a more complex initialization sequence during reset. For example, a description may include a variable that has been mapped to memory (how to do this is discussed in Chapter 9). If every word in the memory needs to be loaded with zeros when the circuit is reset, a multi-cycle initialization sequence is needed. Writing to a memory requires at least one clock cycle, so initializing an entire memory to zeros may take a significant number of clock cycles.

Figure 7-10 shows the general structure of a design that contains both a reset action and an initialization sequence.

```
PROCESS

    init_loop: LOOP

        WAIT UNTIL ( clk'EVENT AND clk = '1');
        EXIT init_loop WHEN ( reset = '1' );

        -- Initialization sequence goes here

        main_loop: LOOP
            -- Normal (non-reset) operation

            WAIT UNTIL ( clk'EVENT AND clk = '1');
            EXIT init_loop WHEN ( reset = '1' );

            -- Algorithm goes here

        END LOOP main_loop;

    END LOOP init_loop;

    -- Reset action goes here

END PROCESS;
```

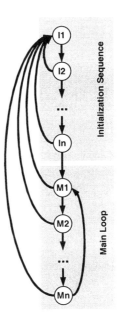

Figure 7-10: General structure of a description with an initialization sequence and corresponding state machine

In this code, an extra loop has been added to contain the initialization sequence. In addition, the EXIT statements that specify the synchronous reset exit the *initialization* loop, *not* the main loop. The reset action now appears after the initialization loop.

The state machine that corresponds to this code is also shown in Figure 7-10. The states that represent the initialization sequence start with the letter "I". Note that the initialization sequence can potentially take many clock cycles to complete. When the initialization is complete, control is transferred to the first state in the main loop. The states that represent the main loop start with the letter "M". When the last state of the main loop is reached, control is passed again to the first state of the main loop.

From every state there is a possible transition to the first state in the initialization sequence. The EXIT statements that represent the synchronous reset imply these transitions.

As long as the reset condition is true, the circuit is held in state I1. The initialization sequence only begins when the reset condition becomes false. Other modules must account for the delay between the reset condition going to false and the start of the main loop when interfacing to a design with an initialization sequence.

Unlike the reset action, the initialization sequence is always performed during behavioral simulation, whether or not a reset is performed. However, since the reset action is after the initialization loop, the reset assignment is still only performed when the reset condition is true.

The code that describes the initialization sequence is scheduled in the same manner as code in the main loop. Thus, unlike the reset action, the code that describes an initialization sequence is not restricted – an initialization sequence can contain multiple WAIT statements, or even additional loops!

6.3.4 Using an Attribute to Specify a Reset Signal

Most behavioral synthesis tools do not require that the reset sequence be explicitly described using EXIT statements in the behavioral description. The EXIT statements can be omitted from the code and the reset signal and phase (active high or active low) can be specified using an attribute or a tool command.

The advantage of omitting the EXIT statements from the behavioral description is that it may make the code appear simpler. To specify a reset in a behavioral description, an EXIT statement must follow *every* WAIT statement. When a behavioral description models the reset behavior, each clock edge is defined using two statements (the WAIT statement and the EXIT statement).

The disadvantage of omitting the EXIT statements is that it is not possible to simulate the reset behavior in the behavioral description, prior to scheduling. But even a design that does not explicitly describe the reset behavior must still contain a reset signal – the reset signal is required to bring the state machine in the synthesized design into a known state.

> *The most reliable design process includes simulation of the reset behavior. This requires that the reset behavior be explicitly modeled in the behavioral description.*

Loops

6.3.5 Specifying an Asynchronous Reset

Some designers prefer to use asynchronous resets instead of synchronous resets. Unlike a synchronous reset that only tests the reset condition on a clock edge, an asynchronous reset immediately changes the state of the circuit when the reset condition becomes true.

This requires a change in the WAIT statement. Under normal operation, the WAIT statement must wait for the next clock edge. However, if the reset condition becomes true while waiting for the clock edge, control must transfer immediately to the reset action. This behavior is accomplished by adding the reset condition to the WAIT statement. The EXIT statement that follows the WAIT statement has the same form as for a synchronous reset. An example of a WAIT / EXIT pair that describes asynchronous behavior is shown in Figure 7-11.

```
WAIT UNTIL (clk'EVENT AND clk = '1') OR ( reset = '1' );
EXIT main_loop WHEN ( reset = '1' );
```

- Wait for a clock edge under normal (non-reset) operation
- Skip to the EXIT statement if the reset condition is true (or becomes true while waiting)
- Jump to the reset action if the reset condition is true

Figure 7-11: Example of a WAIT / EXIT pair that describes asynchronous reset behavior

Aside from the modification to the WAIT statement, a design that describes an asynchronous reset has the same overall structure as a design that describes a synchronous reset. The simulation issues associated with placing the reset action before or after the main loop also apply to asynchronous resets.

6.4 Types of Loops

Loops can be grouped into three categories: infinite loops, WHILE loops, and FOR loops. Unlike infinite loops, which require an EXIT statement to terminate execution of the loop, WHILE loops and FOR loops describe in the loop declaration the conditions under which the loop is terminated.

All three kinds of loops can be used in behavioral descriptions and can use EXIT and NEXT statements to alter the flow of control through the loop.

6.4.1 Infinite Loops

An *infinite loop* is a loop that, *in its declaration*, does not describe any conditions for exiting the loop. An EXIT statement is required to terminate an infinite loop. NEXT statements can also appear in an infinite loop (see Section 6.5).

An infinite loop implies a circular state machine. When the last statement in the loop is reached, execution again continues at the statement at the start of the loop. Branches can be added to the implied state machine using NEXT and EXIT statements.

The beginning of this chapter discussed how an *infinite loop* implies a "circular" state machine. Figure 7-12 shows the state machine that is inferred by an *infinite loop*.

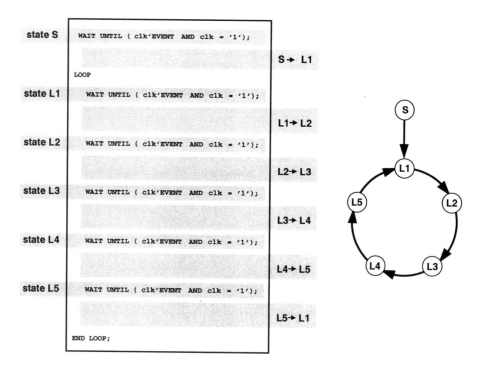

Figure 7-12: An infinite loop implies a "circular" state machine

Loops

6.4.2 WHILE Loops

A WHILE loop is a kind of *unbounded loop*. This means that at compile time, the number of times the loop will be performed is not known. The declaration of a WHILE loop includes an exit condition. When the condition of the loop becomes false, control is transferred to the statement that follows the end of the WHILE loop.

Every WHILE loop can be represented as an infinite loop with an EXIT statement. Often, behavioral synthesis tools translate WHILE loops into infinite loops prior to scheduling. The hardware that is implied by behavioral synthesis is based on the infinite loop, so understanding this translation will simplify the analysis of scheduled results.

Figure 7-13 shows a simple design with a WHILE loop and how that loop can be translated into an infinite loop with an EXIT statement. This design samples the value of the signal **din**. The design then continues to sample this signal until the sum of those values equals or exceeds the value 240.

User Code

```
main_loop : LOOP

    WAIT UNTIL (clk'EVENT AND clk ='1');
    EXIT main_loop WHEN rst ='1';

    sum := din;
    dout <= ( OTHERS => '0' );

    WHILE (sum < 240) LOOP

        WAIT UNTIL (clk'EVENT AND clk ='1');
        EXIT main_loop WHEN rst='1';

        sum := sum + din;

    END LOOP;

    WAIT UNTIL (clk'EVENT AND clk ='1');
    EXIT main_loop WHEN rst ='1';

    dout <= sum;

END LOOP main_loop;
```

Translated Code

```
main_loop : LOOP

    WAIT UNTIL (clk'EVENT AND clk ='1');
    EXIT main_loop WHEN rst ='1';

    sum := din;
    dout <= ( OTHERS => '0' );

    exit_condition := NOT( sum < 240 );

    LOOP
        EXIT WHEN exit_condition;

        WAIT UNTIL (clk'EVENT AND clk ='1');
        EXIT main_loop WHEN rst ='1';

        sum := sum + din;

        exit_condition := NOT( sum < 240 );

    END LOOP;

    WAIT UNTIL (clk'EVENT AND clk ='1');
    EXIT main_loop WHEN rst ='1';

    dout <= sum;

END LOOP main_loop;
```

Figure 7-13: Translation of a WHILE loop into an infinite loop with an EXIT statement

In the translated code the exit condition is calculated in two locations: immediately prior to entering the loop and immediately prior to performing the next iteration of the loop. The duplication of this statement corresponds to the state machine that is constructed for a WHILE loop. Consider how this design simulates. If the value of the input signal **din** is greater than 240 prior to the execution of the WHILE loop, the loop is never executed. This corresponds to a branch in the state machine. However, the exit condition must have been calculated prior to making the branch decision. The exit condition is duplicated outside of the loop so that the exit condition (in this case the "<" operator) can be scheduled in a state before the start of the loop. Of course the calculation of this condition must also be calculated before subsequent iterations of the loop, so the condition is also calculated at the end of the loop. These relationships are illustrated in Figure 7-14.

Translated Code

```
main_loop : LOOP
```
state s1
```
    WAIT UNTIL ( clk'EVENT AND clk ='1');
    EXIT main_loop WHEN rst ='1';

    sum := din;
    dout <= ( OTHERS => '0' );

    exit_condition := NOT( sum < 240 );

    LOOP
        EXIT WHEN exit_condition;
```
state s2
```
        WAIT UNTIL ( clk'EVENT AND clk ='1');
        EXIT main_loop WHEN rst ='1';

        sum := sum + din;

        exit_condition := NOT( sum < 240 );

    END LOOP;
```
state s3
```
    WAIT UNTIL ( clk'EVENT AND clk ='1');
    EXIT main_loop WHEN rst ='1';

    dout <= sum;
END LOOP main_loop;
```

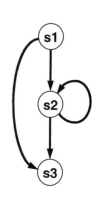

Figure 7-14: Translated WHILE loop and corresponding state machine

There are branches from state **s1** to states **s2** and **s3**. If the exit condition is true, the branch to state **s3** is taken, which bypasses the loop. If the exit condition is false, the branch to state **s2** is taken, entering the loop.

The exit condition is duplicated during the translation of a WHILE loop to an infinite loop so that this decision can be made. It is necessary to calculate the exit condition *prior* to entering the loop. The first occurrence of the exit condition is calculated in state **s1** to decide if the loop is executed or bypassed. The second occurrence of the exit condition is calculated in state **s2** (inside the loop) to decide if the loop has completed.

Loops

6.4.3 FOR Loops

Like WHILE loops, every FOR loop can be represented as an infinite loop. Behavioral synthesis tools also translate FOR loops into infinite loops prior to scheduling.

Figure 7-15 shows a simple design with a FOR loop and how that loop can be translated into an infinite loop. This design samples the value of the signal **din** "n" times.

User Code

```
main_loop : LOOP

    WAIT UNTIL ( clk'EVENT AND clk = '1' );
    EXIT main_loop WHEN rst = '1';

    sum := ( OTHERS => '0' );
    dout <= ( OTHERS => '0' );

    FOR i IN 1 TO n LOOP

        WAIT UNTIL (clk'EVENT AND clk = '1');
        EXIT main_loop WHEN rst = '1';

        sum := sum + din;

    END LOOP;

    WAIT UNTIL ( clk'EVENT AND clk = '1' );
    EXIT main_loop WHEN rst = '1';

    dout <= sum;

END LOOP main_loop;
```

Translated Code

```
main_loop : LOOP

    WAIT UNTIL ( clk'EVENT AND clk = '1' );
    EXIT main_loop WHEN rst = '1';

    sum := din;
    dout <= ( OTHERS => '0' );

    i := 1;
    exit_condition := ( i > n );

    LOOP
        EXIT WHEN exit_condition;

        WAIT UNTIL (clk'EVENT AND clk = '1');
        EXIT main_loop WHEN rst = '1';

        sum := sum + din;

        i := i + 1;
        exit_condition := ( i > n );

    END LOOP;

    WAIT UNTIL (clk'EVENT AND clk = '1');
    EXIT main_loop WHEN rst = '1';

    dout <= sum;

END LOOP main_loop;
```

Figure 7-15: Translation of a FOR loop into an infinite loop

The translation of a FOR loop into an infinite loop is slightly different than the translation for a WHILE loop. In particular, a variable must be created to represent the loop iterator.

The variable that represents the iterator is initialized just before the infinite loop. It is set to the starting value of the iterator, as declared in the FOR loop. In this example, the starting value is 1. Then the exit condition is calculated, similar to a WHILE loop. It is necessary to calculate the exit condition prior to the start of the loop because it is possible that the contents of the loop are not executed at all. For example, if the value of **n** is 0, the loop is not executed.

The exit condition is calculated using the greater-than or less-than operators (> or <), depending upon the direction of the loop (TO and DOWNTO, respectively). These operators are used to account for starting values that are out of range (like in the example when "n" has the value 0).

The translated loop begins with the entire contents of the original loop. At the end of the loop, the next value of the iterator is calculated. The iterator is either incremented or decremented, depending on the direction of the loop (TO or DOWNTO, respectively). Thus, a FOR loop implies either an incrementer or decrementer (in hardware), depending on the direction of the loop.

After the iterator has been incremented (or decremented), its value is tested. If the value is greater-than or less-than (depending upon the direction of the loop) the ending value of the iterator, the loop is complete and is exited. In this example, the ending value of the iterator is "n".

The EXIT statements in the translated code correspond to branching in a state machine in the same manner as for a WHILE loop. Figure 7-16 shows the state machine that corresponds to the translated FOR loop.

Translated Code

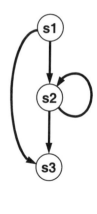

```
main_loop : LOOP
```

state s1
```
    WAIT UNTIL (clk'EVENT AND clk='1');
    EXIT main_loop WHEN rst='1';

    sum := din;
    dout <= ( OTHERS => '0' );

    i := 1;
    exit_condition := ( i > n );

    LOOP
        EXIT WHEN exit_condition;
```

state s2
```
        WAIT UNTIL (clk'EVENT AND clk='1');
        EXIT main_loop WHEN rst='1';

        sum := sum + din;

        i = i + 1;
        exit_condition := ( i > n );

    END LOOP;
```

state s3
```
    WAIT UNTIL (clk'EVENT AND clk='1');
    EXIT main_loop WHEN rst='1';

    dout <= sum;

END LOOP main_loop;
```

Figure 7-16: Translated FOR loop and corresponding state machine

6.5 The NEXT Statement

The NEXT statement is similar to the EXIT statement in that it transfers control when inside a loop. The NEXT statement, however, transfers control to the start of the next iteration of the loop as opposed the statement after the end of the loop. Any statements between the NEXT statement and the end of the loop are not executed.

Loops

When a NEXT statement appears in a FOR loop, the iterator is assigned its next value (incremented or decremented) prior to the transfer of control.

The NEXT statement can be used at the start of a loop to wait until a signal obtains a particular value. For example, a handshaking signal called **start** could be added to the simple algorithm that was introduced in Chapter 2 to indicate that valid data is available at the inputs and that the computation of the algorithm should begin.

Figure 7-17 shows a loop with a NEXT statement and the corresponding state machine.

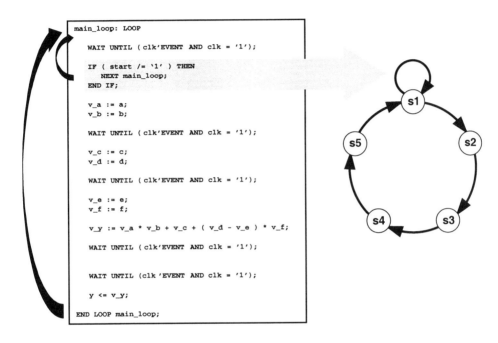

Figure 7-17: NEXT statement used for input handshaking

Note that the condition being tested is `(start /= '1')`. This means that if the value of the signal **start** is not '1', control should be transferred to the next iteration of the loop, in which case the design waits for a clock cycle and tests the value of the signal **start** again. In other words, the value '1' indicates that the input data is valid.

6.6 Scheduling Loops

The following section describes how behavioral synthesis tools schedule loops that appear in behavioral descriptions. The section depicts the corresponding state machines that are automatically constructed by such tools.

A loop is represented by one or more states in the state machine that is synthesized for the design. The general structure of the state machine that is implied by a loop is shown in Figure 7-18.

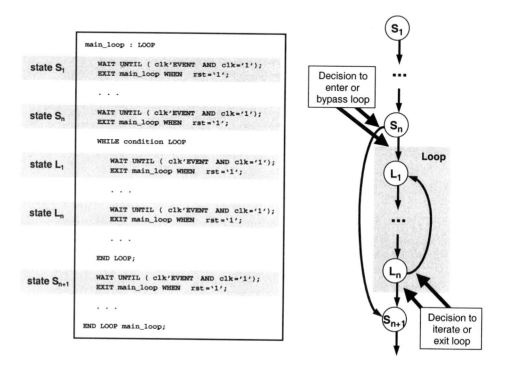

Figure 7-18: The general structure of the state machine implied by a loop

The operations that are used to make the decision to enter a loop or to bypass a loop are scheduled in the state prior to the start of the loop. In the case of a WHILE loop, the exit condition is calculated outside the loop. If the condition is false, the state machine transitions to the first state of the loop (L_1). If the condition is true, the state machine branches to the first state after the end of the loop (S_{n+1}).

Entering a loop corresponds to entering the first state that represents the loop (L_1). When the end of the loop is reached (represented by the state L_n), the exit condition is calculated once again. If the condition is now true, the loop is complete and the state machine exits the loop by transitioning to the first state after the loop (S_{n+1}). If the condition is false, the state machine branches back to the first state of the loop (L_1) to perform another iteration of the loop.

Loops

6.6.1 Clock Cycles and Control Steps

A loop is represented on a Gantt chart in a manner that represents this behavior. Figure 7-19 shows a loop and a Gantt chart that represents a schedule for that loop.

User Code

```
main_loop : LOOP

    WAIT UNTIL (clk'EVENT AND clk ='1');
    EXIT main_loop WHEN rst ='1'

    sum  := ( OTHERS => '0' );
    dout <= ( OTHERS => '0' );

    FOR i IN 1 TO n LOOP

        WAIT UNTIL (clk'EVENT AND clk ='1');
        EXIT main_loop WHEN rst ='1';

        sum := sum + din;

    END LOOP;

    WAIT UNTIL (clk'EVENT AND clk ='1');
    EXIT main_loop WHEN rst ='1';

    dout <= sum;

END LOOP main_loop;
```

Translated Code

```
main_loop : LOOP

    WAIT UNTIL (clk'EVENT AND clk ='1');
    EXIT main_loop WHEN rst ='1'

    sum  := din;
    dout <= ( OTHERS => '0' );

    i := 1;
    exit_condition := ( i > n );

    LOOP
        EXIT WHEN exit_condition;

        WAIT UNTIL (clk'EVENT AND clk ='1');
        EXIT main_loop WHEN rst ='1';

        sum := sum + din;

        i := i + 1;
        exit_condition := ( i > n );

    END LOOP;

    WAIT UNTIL (clk'EVENT AND clk ='1');
    EXIT main_loop WHEN rst ='1';

    dout <= sum;

END LOOP main_loop;
```

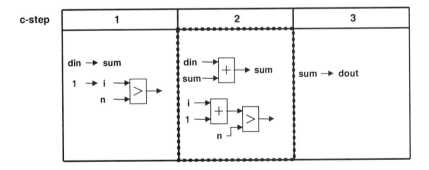

Figure 7-19: FOR loop and corresponding Gantt chart that represents a possible schedule

The loop is delineated from other operations in the chart so that the start and end of the loop are easily recognizable. The loop itself is displayed in a Gantt chart as requiring one or more *control steps* (or *c-steps*).

Until now, this book has referred to *clock cycles* when discussing the scheduling of operations. A more correct term is *c-steps*, or *control steps*. This distinction is made because the term *clock cycles* implies a particular delay.

The source code indicates that this design might take 3 *clock cycles* to compute the value of **dout**, or it might take 20 *clock cycles* to compute – the number of clock cycles required to compute the result depends upon the value of the input signal **n**.

For loops, the Gantt chart indicates the number of clock cycles required to perform *one* iteration of the loop. This information is based on the schedule of the design and is not dependent upon the input data.

Thus, this design is said to be scheduled in 3 *c-steps*, even though it may require many more than 3 *clock cycles* to compute the value of **dout**. It could in fact take less than 3 clock cycles if the loop is not executed at all.

> *The term c-step (or control step) refers to a state in the state machine that is generated during behavioral synthesis. For a loop structure, a Gantt chart shows the control steps (states) for scheduling the operations for a single iteration of the loop. Since a loop may execute multiple times, the actual number of clock cycles required to complete the loop may be significantly different.*

6.6.2 Minimum Loop Execution Time

A loop is represented by one or more states in the state machine that is synthesized for the design. The operations that are used to make the decision to enter a loop or to bypass a loop are scheduled in the state prior to the start of the loop. The loop contents are scheduled onto one or more states. In the last state representing the body of the loop, the decision is made to either exit the loop or perform another iteration of the loop. This general structure was shown in Figure 7-18.

> *If the decision is made to enter a loop, the body of the loop will take at least one cycle to execute. If nothing else, time must be allotted to allow the state machine to determine if the loop should be terminated at the next clock edge. This minimum execution time has broad implications, particularly when trying to synchronize I/O operations using loops.*

Consider the two pieces of code in Figure 7-20 which might be used to wait for a start signal to go high before reading other input signals. Assume that the signals **a** and **b** are valid *only* in the clock cycle when the signal **start** is '1'.

The code on the left was introduced in a previous section. This code uses a NEXT statement to wait until the signal **start** goes to the value '1'. The code waits for a clock cycle and the tests the value of **start**. If the value is '0', the code begins again at the start of the loop, waiting a clock cycle and re-testing the value.

Loops

```
main_loop: LOOP

    WAIT UNTIL ( clk'EVENT AND clk = '1' );

    IF ( start /= '1' ) THEN
        NEXT main_loop;
    END IF;

    v_a := a;
    v_b := b;
    . . .

END LOOP main_loop;
```

```
main_loop: LOOP

    WAIT UNTIL ( clk'EVENT AND clk = '1' );

    WHILE ( start /= '1' ) LOOP
        WAIT UNTIL ( clk'EVENT AND clk = '1' );
    END LOOP;

    WAIT UNTIL ( clk'EVENT AND clk = '1' );

    v_a := a;
    v_b := b;
    . . .

END LOOP main_loop;
```

The code on the right uses a WHILE statement. In this code, the WHILE loop is executed as long as the value of **start** remains at '0'. The loop is exited as soon as the value changes to '1'.

At first glance, these two pieces of code may appear functionally equivalent, but they are not. In the code on the left, the values of **a** and **b** are read in the same clock cycle as when the signal **start** goes to '1'. In the code on the right, the values of **a** and **b** are read in the *next* clock cycle *after* the signal **start** goes to '1'. This is shown in Figure 7-21.

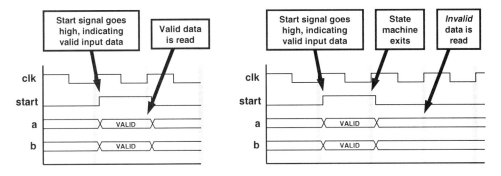

Figure 7-21: Simulation of two code examples used to provide input handshaking

It may seem that the way to "fix" the behavior of the code with the WHILE loop is to remove the WAIT statement that is highlighted in the diagram. But the scheduled design will still behave as if the WAIT statement were there – the signals **a** and **b** will be read in the following cycle.

This behavior is a result of the manner in which the state machines are constructed for loop constructs. Because the body of a loop requires at least one cycle to execute, reading the other input signals cannot occur until the following clock cycle.

When in the loop, the exit condition is calculated to determine whether to exit the loop or perform another iteration of the loop. If the loop is exited, control is transferred to the state immediately following the loop. The values of **a** and **b** are read in this state. Including the WAIT statement in the description makes the behavioral design operate in the same manner as the scheduled design.

6.7 Loop Unrolling

Loop unrolling is a transformation that can be applied to a loop construct. When a loop is unrolled, the statements inside the loop are copied as many times as the loop is being unrolled. When a loop is unrolled, the control structure of the loop is modified to maintain the same functionality – unrolling a loop does not modify the behavior of the loop. There are two types of unrolling: complete unrolling and partial unrolling.

6.7.1 Complete Unrolling

If the number of times a loop will be executed is known during synthesis, the loop can be completely unrolled. FOR loops with constant bounds can be completely unrolled. Since the bounds of the loop are constant the number of iterations is known.

To completely unroll a FOR loop, the statements inside the loop are copied as many times as the loop would be executed. For each copy of the loop statements, the iteration variable is replaced with the appropriate constant for that iteration of the loop. The loop construct is completely replaced by the copied statements. Recall that this is how FOR loops are processed by RTL synthesis tools.

Infinite loops, WHILE loops, and FOR loops with non-constant bounds can not be completely unrolled because the number of times the loop will be performed is data-dependent and thus not known when the design is being synthesized.

Figure 7-22 shows a FOR loop with constant bounds and the equivalent code after the loop has been completely unrolled.

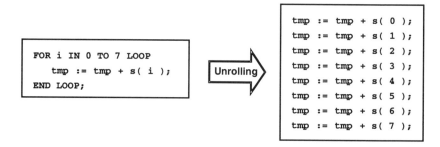

Figure 7-22: Completely unrolling a FOR loop with constant bounds

Some behavioral synthesis tools unroll constant-bound FOR loops by default, others do not. Whatever the default behavior, there is an appropriate mechanism to direct the behavior of the synthesis tool.

It is not always easy to decide whether or not it is best to completely unroll a loop. It is necessary to evaluate the particular loop to make this determination.

> *A loop should only be unrolled if it can reduce the number of cycles required to perform the algorithm.*

Loops

Although not required, it is generally best to unroll certain kinds of loops. When a loop has been used simply as a coding convenience, the loop should be unrolled. Consider the loop in Figure 7-23.

```
FOR i IN 0 TO 6 LOOP
   s( i ) := s( i*2+1 );
END LOOP;
```

Unrolling →

```
s( 0 ) := s(  1 );
s( 1 ) := s(  3 );
s( 2 ) := s(  5 );
s( 3 ) := s(  7 );
s( 4 ) := s(  9 );
s( 5 ) := s( 11 );
s( 6 ) := s( 13 );
```

Figure 7-23: Unrolling a FOR loop with fixed bounds

There is no benefit to keeping this loop rolled. This loop is only used to simplify coding. When the loop is left rolled, 7 clock cycles are required to execute the loop. In addition, hardware is required to control the loop, such as an incrementer and a comparator. Furthermore, the indexed read and indexed write to the array **s** both imply significant hardware (see Section 4-8). But when completely unrolled, the loop only infers wires in the synthesized design!

When a loop contains more substantive operations (such as add and multiply operations), the decision to unroll is more difficult. If a loop is entirely unrolled, there is maximum opportunity to share or chain operations between iterations of the loop. This will reduce the number of clock cycles required to execute the loop. But this usually requires additional hardware.

Consider again the loop shown in Figure 7-22. Assume that the loop is left rolled. If the clock cycle is 50 ns and an add component has a delay of only 3 ns, it will still take 8 clock cycles (400 ns) to execute the loop. The state machine that controls a loop requires at least one clock cycle per iteration of the loop. So when left rolled, the loop implies that one (and only one) addition is performed per clock cycle.

But the number of clock cycles can be reduced if the loop is unrolled. In fact, if the loop is completely unrolled, the loop functionality can be performed in a single clock cycle (50 ns)! Unfortunately, this requires 8 adders in the synthesized hardware.

> *Completely unrolling a loop can increase parallelism but the parallelism comes at a cost of increased hardware. The decision to unroll a loop should be based on the relative cost and benefit of these two factors.*

Constant-bound loops that perform a large number of iterations should not be *completely* unrolled. Unrolling a loop usually results in a greater number of operations that must be scheduled (remember that all the statements inside of the loop are duplicated). Unrolling a loop a large number of times will almost always result in longer scheduling run times in the synthesis tool.

6.7.2 Partial Unrolling

Some behavioral synthesis tools support the *partial unrolling* of loops. To partially unroll a loop, the statements inside the loop are copied as many times as the loop is to be unrolled. The control structure of the loop is appropriately modified to maintain the functionality of the loop.

The loop that was completely unrolled in Figure 7-22 can be partially unrolled. Consider unrolling such a loop 2 times. The translated code that results from the unrolling is shown in Figure 7-24. The translated code shows some obvious "optimization" – the exit condition is set to "false" prior to the start of the loop, as that is the evaluation of the expression "0 > 7".

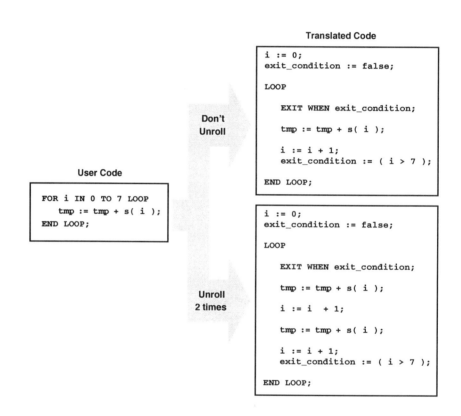

Figure 7-24: Leaving a FOR loop with fixed bounds rolled and unrolling 2 times

There is now an opportunity to create a more parallel implementation. If enough hardware is allocated to perform the operations in the loop in the time of a single clock cycle, the entire loop can be performed in 4 clock cycles instead of 8 clock cycles.

The partial unrolling of the loop does not eliminate the hardware required to perform the indexed read and write to the array **s**. This hardware is not needed only if the indexed references to **s** are always constant. This is only true if the loop is entirely unrolled.

Loops

But the loop functionality can also be achieved in 4 clock cycles by completely unrolling the loop, but limiting the number of adders that can be used in the design. Remember that if the loop is completely unrolled, the loop functionality could be performed in one clock cycle, given 8 adders can be used. If only 2 adders are available, the calculation must be spread over 4 clock cycles. This eliminates the hardware required to index **s** but requires the same number of clock cycles as unrolling the loop 2 times.

The previous example focused on partially unrolling a loop with constant bounds. It is also possible to partially unroll infinite loops, WHILE loops, or FOR loops with non-constant bounds. For these loops, the number of times the loop will execute is data dependent.

When a loop with non-constant bounds is partially unrolled, the exit condition is generated and tested between each copy of the loop contents. This is necessary since the loop may complete at the end of any copy of the loop contents. Figure 7-25 shows an example of unrolling a WHILE loop 3 times.

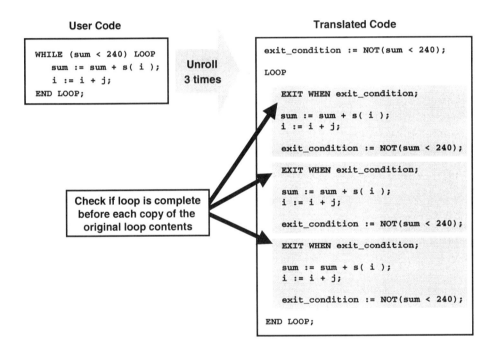

Figure 7-25: Unrolling a WHILE loop 3 times

6.7.3 When to Unroll

It is not always easy to decide whether or not it is best to unroll a loop. It is necessary to understand the design to make this determination.

First, unrolling a loop results in a greater number of operations that must be scheduled. Unrolling the loop may lead to superior results in terms of schedule length or required resources, but unrolling a loop will almost certainly result in longer scheduling run times.

If a loop is entirely *unrolled*, there is maximum opportunity to share or chain operations between iterations of the loop. When entirely unrolled, the loop no longer exists in the design.

If a loop is left *rolled*, there is no opportunity to share or chain operations between different iterations of the loop.

Figure 7-25 shows a loop that is scheduled with the loop left rolled, with the loop completely unrolled, and with the loop unrolled 2 times.

```
FOR i IN 0 TO 7 LOOP
    tmp := tmp + s( i );
END LOOP;
```

	Rolled	Completely Unrolled	Unrolled 2 Times
Schedule			
Resources	1 adder 1 comparitor 1 incrementer indexing logic	4 adders	2 adders 1 comparitor 2 incrementers indexing logic
Latency	8 clock cycles	2 clock cycles	4 clock cycles

Figure 7-25: Simple loop and three schedules (with required resources and latency)

The schedules assume an unlimited number of resources, and a clock cycle large enough to schedule 4 add operations. The resources required to implement the schedules and the number of clock cycles required to perform the loop calculation (latency) are also shown.

6.8 Summary

Loops are fundamental building blocks of behavioral descriptions. Loop structures provide a powerful mechanism for describing the flow of control in an algorithm.

A central feature of behavioral synthesis is the ability to automatically construct a state machine to control the use of resources in a design. The state machine that is implied by a loop structure is also automatically generated.

Unlike RTL synthesis tools, which support only constant-bound FOR loops, behavioral synthesis tools support FOR loops with both constant and non-constant bounds, WHILE loops, and infinite loops. In an RTL design methodology, the use of loop structures is discouraged. Behavioral synthesis methodologies encourage the use of loops to allow designers to describe algorithms at a higher level of abstraction.

Loops can be left rolled or unrolled entirely or partially. Unrolling can dramatically influence the final schedule, but will also impact run-time.

Today's behavioral synthesis tools do not automatically determine when and how loops should be unrolled. This discussion highlights the issues the designer must carefully consider when making these decisions.

Chapter 7

I/O Scheduling Modes

7.1 Overview of Scheduling Modes

For any synthesis tool, be it behavioral or RTL, it is important to understand the relationship between the input description and the generated circuit.

For RTL design, the relationship is quite straightforward. The architecture of the design is explicitly coded in the input description. The input description primarily implies combinational logic or registers, and maps quite easily into gate-level structures. Even though RTL-level optimizations may modify the *micro-architecture* of the design and gate-level optimization may restructure the combinational gates in a design, the overall architecture of an RTL design can be extracted directly from the input description.

For behavioral design, this relationship is not as simple. The architecture of the design is *not* explicitly coded in the description. Behavioral synthesis allows designers to describe circuits at the *architecture-independent* level and to evaluate different architectures that implement that functionality.

Architectures differ by the type and number of components used in the design and the associated data path and controller. But when viewed from the boundary of the design, the only observable difference between architectures is the interface timing, which specifies when input signals are read and when output signals are written.

For some designs, the specification may dictate that behavioral synthesis should not alter the interface timing. For example, a design that processes data based on a network protocol (such as Ethernet or ATM) may need to read and write data in particular clock cycles and at a specified rate that cannot be changed by synthesis.

For other designs, it may be permissible to modify the interface timing as long as the relationships between certain signals are maintained. For example, when the latency (time to compute the result) of a design is not critical, the design may use handshaking signals to interface with external blocks. An "output valid" signal could be used to indicate that the computation is complete. But for this to function correctly, the output valid signal and the output data need to be synchronized to change at the same time.

For yet other designs, the interface timing may be less critical. If the timing specification of a design is flexible, the designer may wish to evaluate the impact of different timing schemes on the generated architectures.

Behavioral synthesis tools can schedule designs in different ways to accommodate these varying requirements.

> *The I/O scheduling mode determines how the timing specified in the behavioral description constrains the interface timing of the circuit generated by behavioral synthesis. When describing a design, it is important to consider the scheduling mode that will be used.*

Behavioral synthesis tools support three different I/O scheduling modes: cycle-fixed, superstate-fixed, and free-floating.

7.2 Cycle-Fixed Scheduling Mode

When a design is scheduled using cycle-fixed scheduling mode (sometimes referred to as fixed mode or fixed I/O mode), the I/O timing of the synthesized design is the same as the I/O timing of the behavioral description.

Fixed mode is best used when the I/O behavior of a design must be rigidly fixed. A characteristic of such designs is that input and output data are read and produced in a fixed manner at a fixed rate.

WAIT statements are used in behavioral models to describe I/O timing. The WAIT statements indicate the clock cycle in which input signals are read and the clock cycle in which output signals are written. The number of WAIT statements in the process determine how long it takes to process the algorithm.

When scheduling in fixed I/O mode, the behavioral code describes not only the functionality of the design but also its timing constraints.

7.2.1 Scheduling a Simple Design

Figure 7-1 shows an algorithm, a timing diagram, and a behavioral description. The behavioral description represents both the functionality of the algorithm and the I/O specification shown in the timing diagram.

The timing diagram shows that the input signals **a** and **b** are only valid in the first clock cycle, the input signals **c** and **d** are only valid in the second clock cycle, and the input signals **e** and **f** are only valid in the third cycle. The diagram also shows that the output signal **y** is registered at the end of the fifth clock cycle and held stable until the next value of **y** is calculated.

Consider scheduling this design using fixed I/O mode. This means that the I/O timing of the scheduled design must match the I/O timing as specified in the behavioral description.

Figure 7-2 shows a set of resources and associated schedule for this design. The schedule is based on a clock period of 10 ns.

Note that even though all of the arithmetic operations (+, -, *) appear in the behavioral description in the same cycle that the output signal **y** is assigned, those operations are not restricted to that cycle, as their operands are *variables* and not signals. Only the I/O operations (signal read and signal write operations) are fixed.

I/O Scheduling Modes

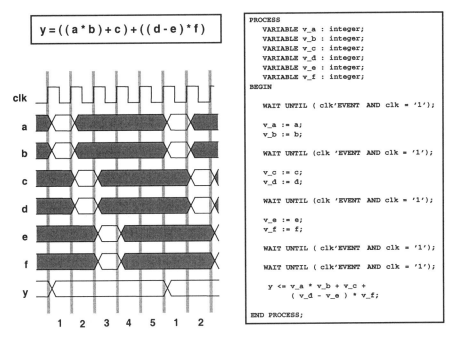

Figure 7-1: Simple algorithm, I/O timing diagram, and behavioral description

Resource	# Available	Delay
Multiplier	2	16 ns
Adder	2	6 ns
Subtractor	1	6 ns

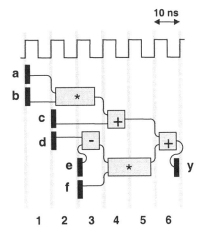

Figure 7-2: Resource allocation and scheduled design based on a 10 ns clock period

Fixed I/O scheduling mode only restricts the interface timing of the design. The input read and output write operations cannot be moved. However, fixed I/O mode does not place restrictions on the scheduling of any other operations in the design. The scheduler is free to place other operations (such as the add, subtract, and multiply operations) into c-steps in a manner that best satisfies the designer's resource constraints.

Consider the scheduled design in Figure 7-2. Because the delay of the multiplier (16 ns) is greater than the clock period (10 ns), the multiply operations must be scheduled as multi-cycle operations, requiring 2 c-steps.

It may seem that the multiplication of the input signals **a** and **b** should begin at the start of the *first* c-step, instead of at the start of the *second* c-step, as it appears in the schedule.

The multiplication begins at the start of the second c-step because of the way that multi-cycle operations are scheduled. Because the multiplier is combinational, but requires 2 clock cycles to compute its result, the inputs to the multiplier must be held stable for the entire time (2 clock cycles) the data is propagated through the multiplier component. But registers can only load data on clock edges. The first opportunity to register the input signals **a** and **b** is at the clock edge between the first and second c-step.

The schedule shows the earliest time that the computation can be completed, based on the specified clock period, the available resources, and the data dependencies implied by the functionality. The earliest that the value of **y** can be computed is in c-step 6. But when scheduling this design in fixed I/O mode, the write to the output signal **y** is fixed in c-step 5.

Thus, this design can not be scheduled in fixed I/O mode with the specified clock period and resources. Behavioral synthesis tools will produce an error when attempting to schedule this design in fixed I/O mode.

This illustrates one of the disadvantages of using fixed I/O scheduling mode. By fixing the c-steps in which input signals and read and output signals are written, fixed I/O mode greatly restricts the architectures that can be constructed.

> *Cycle-Fixed I/O mode is the most restrictive I/O mode for scheduling. If the output values can not be computed in the c-steps specified in the behavioral description with the resources provided, scheduling will fail.*

Now consider scheduling this design using the same resources, but with a clock period of 12 ns. Figure 7-3 shows the schedule for the design.

It is now possible to schedule this design in fixed I/O mode. This is because the longer clock period allows a multiplier and an adder to be *chained* together. The delay through both components is 22 ns, which is less than the delay of 2 clock cycles (24 ns). Thus, with a 12 ns clock period, this design can be successfully scheduled in fixed I/O mode with the resources provided.

Even in fixed I/O mode, there are still many constraints that the designer can modify to explore alternate architectures. As in this example, the clock period can be varied. The number and type of resources can be controlled. Different loop unrolling, pipelining, and memory mapping techniques can also be explored.

I/O Scheduling Modes

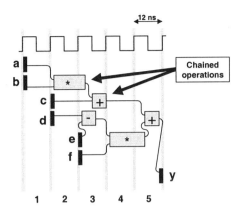

Figure 7-3: Scheduled design based on a 12 ns clock period

7.2.2 Testing Designs Scheduled in Fixed I/O Mode

When a design is scheduled using fixed I/O mode, the resulting circuit can be easily tested. Since the I/O timing of the synthesized circuit is known, only a simple test bench is necessary. The test bench can simply apply input stimulus in the correct clock cycles, wait the appropriate number of cycles while the computation is being performed, then test the output values.

Since the I/O timing of the design is not modified in any way during scheduling, only one test bench is needed for both the source and synthesized designs. This greatly simplifies the testing strategy for the design. A design process that requires different test benches for each stage in the process is likely to be error-prone.

> *Designs scheduled using fixed I/O mode require only a simple test bench that can be easily used for both the source and synthesized designs.*

Test benches are described in detail in Chapter 12.

7.2.3 Scheduling Conditional Branches in Fixed I/O Mode

The previous example shows the scheduling of straight-line code. Now consider scheduling conditional statements (IF and CASE statements).

Behavioral synthesis tools can schedule conditional statements in two different manners, depending upon the design requirements. The branching in the conditional can either be implemented as multiplexer logic in the data path portion of the design, or as branching in the control portion of the design.

When conditional branching is implemented in the data path, every possible branch of the conditional is computed in parallel, then a multiplexer structure selects the appropriate data. This is similar to how conditional statements are represented in hardware by RTL synthesis tools.

Unlike RTL synthesis tools, behavioral synthesis tools can schedule the operations associated with a conditional statement into multiple clock cycles. But in order to represent the conditional branching entirely in the data path portion of the design, every branch is scheduled in the same number of clock cycles.

Behavioral synthesis schedules conditional branches in this manner when there are no WAIT statements inside the branches of the conditional.

Consider the behavioral code in Figure 7-4. Neither the THEN clause nor the ELSE clause contain any WAIT statements.

```
state s0    WAIT UNTIL clk'EVENT AND clk='1';
            IF ( sel = '1' ) THEN
                y <= a + b;
            ELSE
                y <= a - b;
            END IF;
state s1    WAIT UNTIL clk'EVENT AND clk='1';
```

Figure 7-4: Conditional statement with branches of equal length

Figure 7-5 shows the hardware that is generated for this code. The entire conditional is evaluated when the controller is in state **s0**. The branching in the IF statement is represented as a multiplexer in the data path. A single state in the controller affects when the output register is loaded.

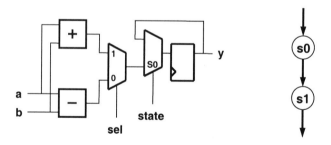

Figure 7-5: Implementation for a conditional statement with branches of equal length

> When the branches of a conditional statement do not contain any WAIT statements, the branching is represented in the data path. The entire conditional can take multiple clock cycles, but each branch takes the same number of clock cycles to execute.

I/O Scheduling Modes

WAIT statements can be placed in the branches of a conditional when the branches should take different numbers of clock cycles. Each branch can contain a different number of WAIT statements to correctly represent the desired latency.

But when each branch can take a different numbers of clock cycles to compute, it is not possible to represent the conditional entirely in the data path. The number of clock cycles that are required to process the conditional depends upon the branch that is selected. This affects the state machine for the design.

When a behavioral description contains such a conditional, the branching is represented in the controller. Each branch is represented by one or more states.

Figure 7-6 shows an IF statement with branches of different length and the corresponding state machine.

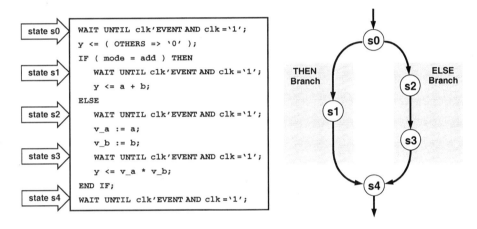

Figure 7-6: IF statement with branches of differing length

In this example, the THEN clause contains a single WAIT statement and thus is represented by one state in the controller. The ELSE clause contains two WAIT statements and is represented by two states.

The condition, (**"mode = add"**) is evaluated in state **s0**. If the expression evaluates to true, control is transferred to state **s1**. If the expression is false, control is transferred to state **s2**. Both branches transfer control to state **s4** when they have completed.

> *When the branches of a conditional statement contain WAIT statements, the branching is represented in the controller. Each branch can take a different number of clock cycles to execute and are represented by one or more states.*

The way in which conditionals with uneven branches are scheduled places restrictions on the way in which they can be described in behavioral code.

In this example, the THEN clause is represented by a single state (**s1**). In fixed I/O mode, it must be possible to schedule every operation in that branch of the conditional in a single c-step. This implies that the delay through the adder component that is used for this operation must be less than one clock cycle. If the add operation required multiple clock cycles, scheduling this design in fixed I/O mode would fail.

In state **s0**, a decision must be made about which branch should be entered. Each branch is represented in the controller by one or more states. Depending upon the value of the condition, control is transferred to the first state that represents the appropriate branch (either **s1** or **s2**). When the branch has been processed, control is then transferred to the state that follows the conditional statement (**s4**).

Each branch is represented in the state machine by at least one state, and thus each branch takes at least one clock cycle to execute. This means that in the behavioral description, if *any* branch contains one or more WAIT statements, then *every* branch that contains operations must contain at least one WAIT statement.

Note that the states that represent the branches are separate from other states in the controller. Thus, the statements in the behavioral description that precede the conditional statement or follow the conditional statement must be separated from the code in each branch by WAIT statements. In the previous example, the WAIT statements that correspond to the states **s1**, **s2**, **s3**, and **s4** isolate the statements in the branches.

In order to isolate the assignments that occur before and after a conditional, there are three conditions that must be met when describing a conditional with unequal branch lengths in fixed I/O mode:

- Operations that occur *prior* to the start of the conditional must be separated from operations inside each branch by at least one WAIT statement.

- Operations inside each branch of the conditional must be separated from operations that occur *after* the end of the conditional by at least one WAIT statement.

- Operations that occur *prior* to the start of the conditional must be separated from operations that occur *after* the end of the conditional by at least one WAIT statement.

Figure 7-7 shows examples of conditional statements that are not legal in fixed I/O mode. These examples violate one of the previously-stated requirements.

I/O Scheduling Modes

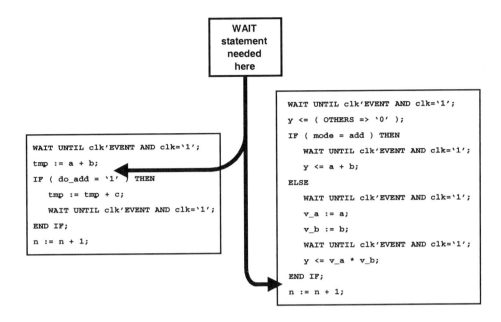

Figure 7-7: Conditional statements that are illegal in fixed I/O mode

CASE statements can also have unequal branch lengths and result in more complex branching in the state machine. Figure 7-8 shows an example of a CASE statement with 4 possible selections.

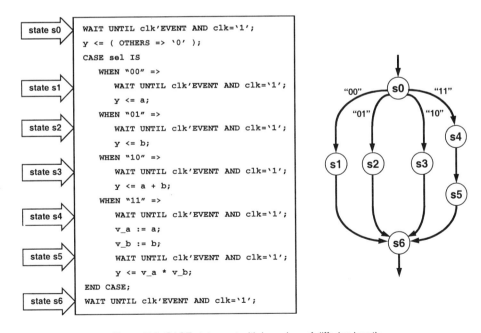

Figure 7-8: CASE statement with branches of differing length

7.2.4 Scheduling Loops in Fixed I/O Mode

Recall how loops are represented in hardware, as discussed in Chapter 6.

When hardware is generated for a loop, the loop is represented in the controller by one or more states. In the state prior to the start of a loop, a decision is made to either enter the loop or to bypass the loop. In the last state of the loop, a decision is made to either iterate on the loop or to exit the loop.

The states that represent the loop are separate from other states in the controller. Thus, the statements in the behavioral description that precede the loop or follow the loop must be separated from the loop by WAIT statements.

In order to isolate the assignments that occur before and after a loop, there are three conditions that must be met when describing a loop in fixed I/O mode:

- Operations that occur *before* the loop must be separated from operations inside the loop by at least one WAIT statement.

- Operations inside the loop must be separated from operations that occur *after* the loop by at least one WAIT statement.

- When iterating on a loop (because a NEXT statement or the end of the loop has been reached), operations that occur just before iterating on the loop must be separated from operations that occur at the *start* of the loop by at least one WAIT statement.

- Operations that occur *before* the loop must be separated from operations that occur *after* the loop by at least one WAIT statement.

Usually these conditions can be satisfied by including a WAIT statement as the first statement inside the loop and as the first statement after the loop. Consider the rather complicated loop in Figure 7-9.

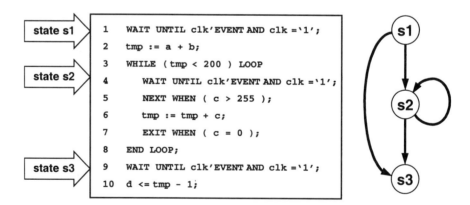

Figure 7-9: WHILE loop with WAIT statements at the start of and after the loop

I/O Scheduling Modes

There are many possible paths through the WHILE loop that must be considered:

- If the loop is *bypassed*, the operations in line 2 and the operations in line 10 are separated by the WAIT statement at line 9.
- If the loop is *entered*, the operations in line 2 and the operations in line 5 are separated by the WAIT statement at line 4.
- If the condition associated with the NEXT statement at line 5 is true, the operations in the current iteration of the loop and the operations in the next iteration of the loop are separated by the WAIT statement at line 4.
- Similarly, if the loop *iterates* because line 8 is reached, the operations in the current iteration of the loop and the operations in the next iteration of the loop are separated by the WAIT statement at line 4.
- If the loop is *exited* from the EXIT statement in line 7, the operations inside the loop and the operations outside the loop are separated by the WAIT statement at line 9.
- Finally, if the loop is *exited* when the condition in the WHILE loop becomes false, the operations inside the loop and the operations outside the loop are separated by the WAIT statement at line 9.

It is necessary that all of these conditions are met to be able to schedule this example in fixed I/O mode.

Some behavioral synthesis tools allow an exception to the rules requiring the separation of operations. Consider the two examples in Figure 7-10.

It appears that these examples should be illegal for fixed I/O mode, because they do not contain WAIT statements in all the locations required to satisfy the rules for fixed I/O mode.

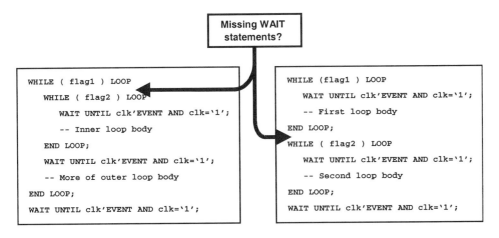

Figure 7-10: WHILE loops that appear to have missing WAIT statements

Some behavioral synthesis tools allow these examples to be scheduled in fixed I/O mode because the conditions that control the WHILE loops are so simple. In these examples, the conditions are boolean flags whose values require little or no time to compute.

Such "negligible-delay" operations can be evaluated on the transition between states. It is not necessary to allocate an entire state simply to test the value of a boolean. The result is a slightly more complex controller, but the benefit is the elimination of an entire c-step.

"Negligible-delay" operations include variable reads, signal reads, variable writes, and simple comparisons to single-bit signals and variables. Note that these operations do add a small amount of delay to the controller but for a savings of an entire c-step. All other operations that do not fall into this category must be associated with a state.

To understand the benefit of this feature, consider the example in Figure 7-11. The figure shows the state machine for two sequential WHILE loops, each with *complex* operations as conditions.

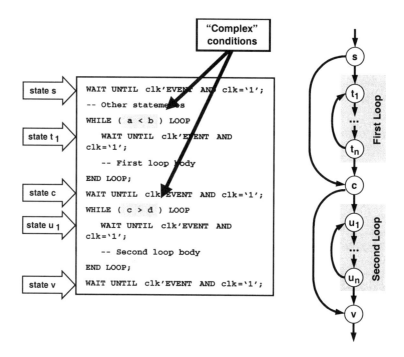

Figure 7-11: Sequential WHILE loops and associated state machine

The rules of fixed I/O mode dictate that the behavioral description include a WAIT statement between the two loops. This WAIT statement corresponds to state **c**, which is required to provide time to calculate the exit condition (**c > d**). The state machine shows that control must always pass through state **c**. This means that it takes 2 clock cycles to just bypass the loops!

I/O Scheduling Modes

Figure 7-12 shows the state machines for similar sequential WHILE loops, but these loops have only simple boolean variables as conditions.

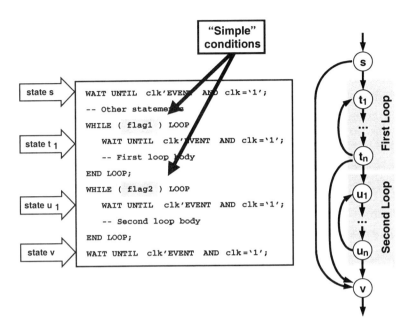

Figure 7-12: Sequential WHILE loops and associated state machine

In this example, the behavioral description does *not* require a WAIT statement between the two loops. The simple conditions can be associated with the *transitions* between states and not in the states themselves. This eliminates the state that was called **c** in the previous example. Without the intermediate state, control can transfer directly from the state before the loops to the state after the loops. Thus, for this example it takes only a single clock cycle to bypass both loops.

7.2.5 Advantages and Disadvantages of Fixed I/O Mode

The following lists describe some of the advantages and disadvantages associated with fixed I/O mode.

Some advantages of fixed I/O mode are:

- The I/O timing of synthesized design matches the behavioral description.

    The I/O timing of the design is not modified in any way during synthesis.

- The I/O timing of the behavioral description is easier to understand.

    The I/O timing is explicitly declared in the behavioral description. It is very easy to translate a timing diagram into a behavioral description or a behavioral description into a timing diagram.

- Designs require only a simple test bench that can be easily used on both the behavioral description and the synthesized designs.

    The I/O timing of the design is not modified in any way during synthesis. A test bench can simply apply input stimulus in the correct clock cycles, wait the appropriate number of cycles while the computation is being performed, then test the output values. The test bench is very simple and is the same for both the source and synthesized designs.

Some disadvantages of fixed I/O mode are:

- Fixed I/O mode is the most restrictive I/O mode.

    Fixed I/O mode is the most restrictive I/O mode for scheduling. If the output values can not be computed in the specified cycles with the resources provided, scheduling will fail (as in the first scheduling example).

- Changes in I/O timing require modifications to the behavioral description.

    The I/O timing is explicitly declared in the behavioral description. Any changes to that timing must be reflected in the behavioral description.

When describing a design that will be scheduled using fixed I/O mode, it is necessary to consider the possible implementations of the design. By fixing the I/O behavior, architectural exploration may be limited such that the description in fact implies a certain architecture. In this way, designs described for fixed I/O mode are closest to RTL descriptions.

7.3 Superstate-Fixed Scheduling Mode

One of the disadvantages of fixed I/O mode is that it does not allow the latency of the design to be modified by synthesis. Often the latency of the design is an aspect of exploration that the designer wishes to consider.

Recall the schedule in the previous section that used a 10 ns clock. It was not possible to schedule this design in fixed I/O mode because the earliest the signal **dout** could be written was in c-step 6, while the behavioral description dictated that it *must* be done in c-step 5. But for some applications, an architecture that generated the result in c-step 6 might be acceptable. Fixed I/O mode would fail when attempting to schedule this design.

Superstate-Fixed mode (sometimes referred to as Superstate mode) provides another way of interpreting I/O read and write operations in a behavioral description to help address this issue.

Unlike fixed I/O mode, in which the time between consecutive WAIT statements represents a *single* clock cycle, in superstate mode, consecutive WAIT statements can represent *one or more* clock cycles. In superstate mode, WAIT statements in the behavioral description define the boundaries of a *superstate*.

I/O Scheduling Modes

The following rules apply to superstates:

- A superstate can represent one or more c-steps in the synthesized design.

  This rule allows the latency of the synthesized design to differ from the latency of the behavioral description. Clock cycles in the source description can be "stretched" into multiple clock cycles in the scheduled design.

- I/O operations (signal reads and signal writes) can not cross a superstate boundary.

  I/O operations are constrained to the superstate in which they are written and behavioral synthesis is not allowed to move them across superstate boundaries. Synthesis can add as many clock cycles to a superstate as is necessary to schedule the operations in the design, but the <u>ordering</u> of the I/O reads and writes is not changed by synthesis.

- If a superstate is represented by more than one c-step, the scheduler assumes that an input can be read in any c-step within the superstate.

  Superstate I/O mode makes this assumption to provide greater flexibility during scheduling. But this places the burden on the user to ensure that the input data is valid when it is read. One way to ensure that only valid data is read is to make the input data that is read in a particular superstate stable throughout every c-step in that superstate. Alternately, the user can examine the schedule of the synthesized design and then modify the external modules to produce data in the particular c-steps in which that data is read in the synthesized design.

- If a superstate is represented by more than one c-step, output signals are only written to in the last c-step within the superstate.

  If more than one output signal is written to in a superstate, behavioral synthesis assumes that the designer intended these signals to change in the same c-step. Consider an "output-valid" signal that is used to indicate that a computation is complete. For the design to function correctly, the output-valid signal and the output data need to be synchronized to change at the same time. If these signals are assigned in the same superstate in the behavioral description, this rule will ensure that they change in the same c-step, even if the superstate is represented as multiple c-steps in the synthesized design.

As in fixed I/O mode, variable assignments and other operations (such as add or multiply operations) are not fixed in superstate mode and can be moved by the scheduler.

Superstate mode is best used when the sequence of I/O operations is important but the exact cycle-to-cycle correspondence between simulation of the behavioral description and the synthesized design is not. This mode is ideal when a handshaking protocol is used to synchronize input and output events, but the number of clock cycles required to perform the calculation is either unknown or may vary depending upon the number and type of available resources. (Handshaking is discussed in Chapter 11).

7.3.1 Scheduling a Simple Design

Recall the example discussed in Section 7.2.1 that could not be scheduled in fixed I/O mode with a clock period of 10 ns. Figure 7-13 shows the algorithm, a timing diagram, and a behavioral description.

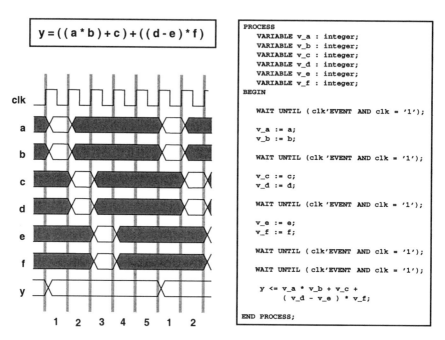

Figure 7-13: Simple algorithm, I/O timing diagram, and behavioral description

The minimum number of cycles required to schedule the design in fixed I/O mode was 6 cycles. This violated the fixed I/O mode constraint derived from the I/O timing of the behavioral description. The behavioral description indicated that the design had to be scheduled in 5 cycles. In fixed I/O mode, it was necessary to increase the cycle time to 12 ns to schedule the design.

Now consider scheduling this design in superstate I/O mode. In superstate I/O mode, the I/O constraints are more relaxed: only the ordering of I/O reads and writes is restricted; cycles can be added between WAIT statements as are needed to schedule the design.

Figure 7-14 shows this design scheduled in superstate I/O mode.

The design can be scheduled in superstate mode, since the last superstate can be "stretched" into 2 clock cycles (as shown in the Gantt chart), without altering the order of the I/O operations.

I/O Scheduling Modes

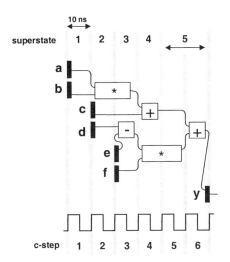

Figure 7-14: Simple algorithm scheduled in superstate I/O mode

The following table provides a quick comparison between superstate I/O mode and fixed I/O mode:

Fixed I/O Mode	**Superstate I/O Mode**
In fixed I/O mode, each WAIT statement represents a clock edge; the delay between two consecutive WAIT statements is *one* clock cycle.	In superstate I/O mode, each WAIT statement represents the boundary of a *superstate*. The delay between consecutive WAIT statements is *one or more* clock cycles.
In fixed I/O mode, the behavioral description specifies the I/O timing and latency required by the designer; these cannot be modified during architectural exploration.	In superstate I/O mode, the behavioral description specifies the I/O ordering and *minimum* (or sometimes desired) latency required by the designer; exploration is extended to include the latency of the design, but without disturbing the ordering of I/O operations.

7.3.2 Testing Designs Scheduled in Superstate I/O Mode

Note that the design represented by the schedule shown in Figure 7-14 will simulate differently than the behavioral description from which it was created. In the behavioral description, the signal **y** is written in c-step 5, but in the synthesized design it is written in c-step 6. Also, the latency between the two designs is different. The behavioral description samples new data every 5 clock cycles, but the synthesized design samples data every 6 clock cycles.

These differences will cause problems for testing. Unlike in fixed I/O mode, the test bench cannot simply apply input stimulus, wait some number of clock cycles, then test the output values -- the number of cycles to wait is different between the source and synthesized designs.

One possible solution to this problem is to have two test benches: one for the behavioral description and one for the synthesized design. But this greatly complicates the testing strategy for the design. A design process that requires different test benches for each stage in the process is likely to be error-prone.

Another solution to this problem is to introduce handshaking signals into the design to synchronize events with external blocks.

Consider a slightly modified algorithm that includes an output signal called **valid** that is used to indicate that the computation is complete. Figure 7-15 shows the modified algorithm, timing diagram, and behavioral description.

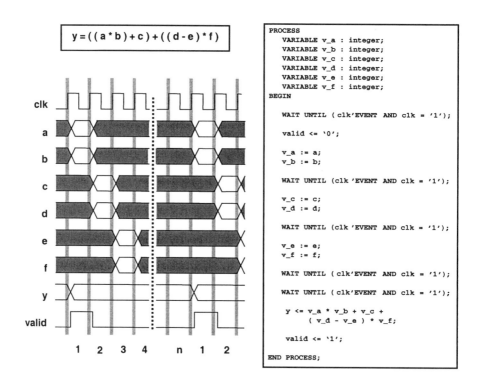

Figure 7-15: Simple algorithm with output handshaking, timing diagram, and behavioral description

Just like in the previous example, the timing diagram shows that the input signals **a** and **b** are only valid in the first clock cycle, the input signals **c** and **d** are only valid in the second clock cycle, and the input signals **e** and **f** are only valid in the third cycle. But the number of cycles required to calculate the result is not specified. The diagram does, however, show that the signal **valid** goes high when the output signal **y** is valid, then goes back low while the next result is being calculated.

I/O Scheduling Modes

Now consider scheduling this modified design in superstate I/O mode. The schedule is shown in Figure 7-16.

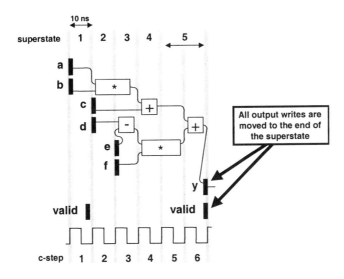

Figure 7-16: Simple algorithm with output handshaking signal scheduled in superstate I/O mode

In this schedule, the signal **y** is written in c-step 6, which is at the end of the superstate that is represented by c-steps 5 and 6. Even though the write to the signal **valid** could be scheduled earlier (it doesn't have any data dependencies), superstate I/O mode ensures that the write is placed in the last c-step of the superstate. This results in the desired behavior. Even though the latency of the behavioral description and the scheduled design are different, the signals **y** and **valid** are written to in the same c-step in both.

When a design with handshaking signals is scheduled using superstate I/O mode the resulting circuit can be easily tested. Even though the latency of the synthesized circuit can be greater than that of the behavioral description, superstate mode ensures that I/O write operations are always placed in the last c-step of a superstate. This allows the designer to synchronize data output signals with output handshaking signals. The test bench can apply input stimulus, wait until the output handshaking signal goes active (high in this example), then test the output values. This alleviates the need for multiple test benches.

> *When scheduling using superstate I/O mode, designs can be written using handshaking signals to synchronize I/O events. This facilitates the use of a simple test bench that can be used on both the source and synthesized designs.*

Handshaking is described in detail in Chapter 11.

Test benches are described in detail in Chapter 12.

7.3.3 Scheduling Conditional Branches in Superstate I/O Mode

Conditional statements (IF and CASE statements) will be scheduled in two different manners in superstate I/O mode depending on the presence of WAIT statements in the branches of the condition.

If there are *no* WAIT statements in any branch of the conditional, then the branching logic will be represented in the data path, as it is with fixed I/O mode when no WAIT statements are used.

But unlike fixed I/O mode, in superstate I/O mode the number of cycles required to process a conditional statement can be extended by the behavioral synthesis tool with the addition of c-steps into a superstate. However, after scheduling, each branch in the conditional will be scheduled in the same number of c-steps.

Consider scheduling the design shown in Figure 7-17 in superstate I/O mode.

```
WAIT UNTIL clk'EVENT AND clk ='1';

IF ( cond = '1' ) THEN
    d <= a + b + c;
ELSE
    d <= a - b;
END LOOP;

WAIT UNTIL clk'EVENT AND clk ='1';
```

Figure 7-17: IF statement without WAIT statements in branches

Because there are no WAIT statements in the branches of the conditional, the branches will have equal length after scheduling. Assuming only one add-sub component is available, and that component can calculate a result in less than one c-step, the conditional will be scheduled as shown in Figure 7-18.

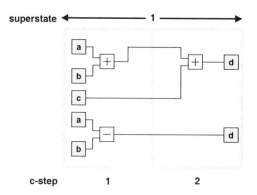

Figure 7-18: Scheduling an IF statement without WAIT statements in branches

I/O Scheduling Modes

Note that the first add operation in the THEN clause and the subtract operation in the ELSE clause can both be scheduled in the same c-step even though only one add-sub component is available. This is because the operations are mutually exclusive – the operations can never both be active at the same time.

The signal **d** is written to in the second c-step, regardless of the value of the input signal **cond**. This is consistent with the rule of superstate I/O mode, which places all write operations into the last c-step of a superstate. Each branch of the conditional requires 2 clock cycles to calculate the value of **d**.

Conditionals are scheduled very differently when there *are* WAIT statements in the branches of the conditional. In this case, then the branching logic will be represented in the controller, as it is with fixed I/O mode when WAIT statements are used.

In this case, each *branch* represents a superstate. This means that every branch in the conditional must contain at least one WAIT statement (in order to be defined as a superstate) -- if one branch contains a WAIT statement, every branch must contain a WAIT statement.

Since each branch is treated as a superstate, the number of cycles required to process each branch can be extended by the addition of c-steps into a superstate. After scheduling, each branch in the conditional may be scheduled in different numbers of c-steps.

Consider scheduling the design shown in Figure 7-19 in superstate I/O mode.

```
WAIT UNTIL clk'EVENT AND clk='1';

IF ( cond = '1' ) THEN
   WAIT UNTIL clk'EVENT AND clk='1';
   WAIT UNTIL clk'EVENT AND clk='1';
   d <= ( a + b ) * c;
ELSE
   WAIT UNTIL clk'EVENT AND clk='1';
   d <= a - b;
END IF;

WAIT UNTIL clk'EVENT AND clk='1';
```

Figure 7-19: IF statement with WAIT statements in branches

Because the THEN clause contains two WAIT statements, it will require *at least* two c-steps to schedule. Similarly, the ELSE clause contains one WAIT statement and will thus require at least one c-step to schedule. Since the WAIT statements represent the boundaries of superstates, the actual number of c-steps could be greater.

Assume that one add-sub component is available and that the component can calculate a result in less than one c-step. Also assume that one multiply component is available, but that it requires two c-steps to calculate a result. The conditional will be scheduled in superstate I/O mode as shown in Figure 7-20.

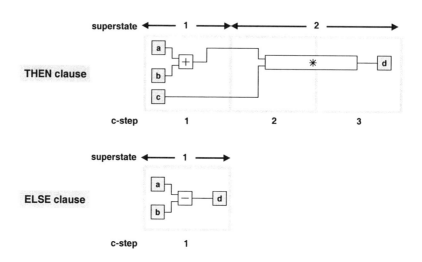

Figure 7-20: Scheduling an IF statement with WAIT statements in branches

When a conditional statement contains WAIT statements, each branch is also represented in the controller as separate sets of states, similar to fixed I/O mode. For such conditionals, three conditions must be met in superstate I/O mode. These are the same conditions that must be met when describing a conditional with unequal branch lengths in fixed I/O mode:

- Operations that occur *prior* to the start of the conditional must be separated from operations inside each branch by at least one WAIT statement.

- Operations inside each branch of the conditional must be separated from operations that occur *after* the end of the conditional by at least one WAIT statement.

- Operations that occur *prior* to the start of the conditional must be separated from operations that occur *after* the end of the conditional by at least one WAIT statement.

These rules also isolate the superstates that precede and follow the conditional from the conditional itself. When a branch of a conditional is scheduled as a superstate, the superstate cannot extend beyond the boundaries of the conditional statement.

7.3.4 Scheduling Loops in Superstate I/O Mode

Previous discussions have shown how loops are represented by one or more states in the controller. The operation of a loop can be thought of as being "isolated" from the operation of the rest of the design.

This has implications for scheduling. Assignments to signals (signal write operations) that occur before a loop in a behavioral description are scheduled prior to the start of the loop. Similarly, assignments to signals that occur inside a loop are scheduled prior to the end of the loop.

But superstate I/O mode dictates that signal write operations be placed in the *last* c-step of a superstate. To avoid contradictory requirements, superstates that contain assignments to signals must stop at loop boundaries. If they did not, the rules of superstate I/O mode would attempt to move the write operations into (or even past) the loop. This means that a superstate cannot extend into a loop, nor can it extend beyond the end of the loop.

This requirement translates into restrictions on the behavioral description. These rules are:

- Signal write operations cannot be moved *into* a loop.

    Signal writes that *precede* a loop must be separated from the *body* of the loop by at least one WAIT statement.

- Signal write operations cannot be moved *over* a loop.

    Signal writes that *precede* a loop must be separated from statements *after* the loop by at least one WAIT statement.

- Signal write operations cannot be moved *out* of a loop.

    Signal writes *inside* a loop must be separated from statements *after* the loop by at least one WAIT statement. EXIT statements must be considered when satisfying this rule.

- Signal write operations cannot be moved from the *bottom* of a loop to the *top* of the loop.

    Signal writes that occur after the *last* WAIT statement (i.e. in the last superstate) in the body of a loop must be separated from operations at the start (i.e. before the first WAIT statement) of the loop by at least one WAIT statement.

These rules are very similar to the restrictions associated with loops for fixed I/O mode. But note that these rules are based on the feature of superstate I/O mode that places all signal writes in the last c-step of the superstate. If there are only writes to variables, it is not necessary to include the WAIT statements.

Figure 7-21 shows examples of behavioral descriptions that violate each of the rules associated with loops and superstate I/O mode and the ways in which the rules can be fixed.

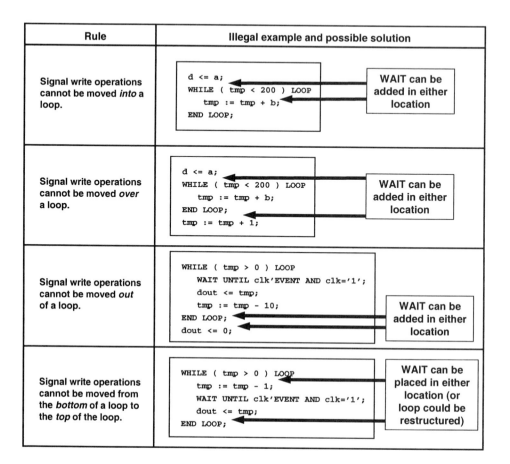

Figure 7-21: Behavioral descriptions with loops that violate superstate I/O mode rules

7.3.5 Advantages and Disadvantages of Superstate-Fixed I/O mode

The following lists describe some of the advantages and disadvantages associated with superstate I/O mode.

Some advantages of superstate I/O mode are:

- The I/O timing of the synthesized design is similar to the source.

    The *ordering* of I/O operations is not modified during scheduling. However, the latency of the design may be longer than the source, if superstates are expanded into multiple c-steps.

I/O Scheduling Modes

- The I/O timing of the source is easy to write and understand.

    Even though superstates can represent multiple c-steps, the I/O timing can still be viewed as being explicitly declared in the behavioral description. It is very easy to translate a timing diagram into a behavioral description and a behavioral description into a timing diagram.

- If handshaking is used, designs require only a simple test bench that can be easily used on both source and synthesized designs.

    If the design uses handshaking signals, then a test bench can simply apply input stimulus, wait until an "output-valid" signal is active, then test the output values. The test bench can still be relatively simple and would be the same for both the source and synthesized designs.

- Superstate I/O mode is less restrictive than fixed I/O mode and provides for more exploration opportunity

    Superstate I/O mode is less restrictive than fixed I/O mode for scheduling. If the output values can not be computed in the number of available cycles with the resources provided, clock cycles can be added to superstates. However, the ordering of I/O operations will not be modified.

Some disadvantages of superstate I/O mode are:

- Changes in I/O timing require source code modifications.

    Even though superstates can represent multiple clock cycles, the I/O timing can still be viewed as being explicitly declared in the behavioral description. Any changes to that timing must be reflected in the source code. Modifications to timing can also be made with additional constraints, but should ultimately be incorporated into the behavioral description.

- It may be necessary to specify additional constraints.

    Superstates can represent multiple clock cycles. If it is important that two read operations occur in consecutive clock cycles, but the scheduler wishes to "stretch" the superstate between them, it would be necessary to add a constraint that forces the delay between the operations to be 1 clock cycle.

7.4 Free-Floating Scheduling Mode

Free-Floating I/O mode (usually referred to as free mode or free I/O mode) is the least restrictive scheduling mode used by behavioral synthesis tools. In free mode, synthesis is totally free to add or delete clock cycles and to move I/O operations. This provides the tool with maximum freedom to explore architectural alternatives. However, since free I/O mode may move I/O operations, the I/O protocol of the scheduled design may not match the I/O protocol of the behavioral description.

Free mode is best used when the sequencing of I/O operations is not important or has not yet been defined, or to explore architectural alternatives independent of the limitations that may be imposed by I/O timing requirements. When used in this manner, the results of scheduling in free mode may influence how the I/O timing of a design is ultimately defined. In this mode the designer is exploring architectural alternatives with the fewest constraints.

7.4.1 Scheduling a Simple Design

Figure 7-22 shows an algorithm, a timing diagram, and a behavioral description. The behavioral description represents both the functionality of the algorithm and the I/O specification shown in the timing diagram.

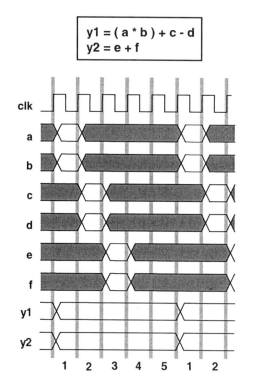

Figure 7-22: Simple algorithm, I/O timing diagram, and behavioral description

I/O Scheduling Modes

Consider scheduling this design using free I/O mode. This means that scheduling can add or delete clock cycles or move I/O operations. Assuming a multiply component that requires one and a half clock cycles and an adder and subtractor component that each take less than half a clock cycle, a schedule is shown in Figure 7-23.

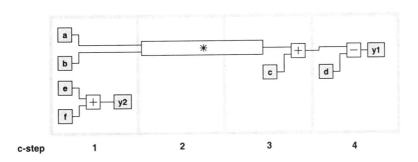

Figure 7-23: Scheduling a simple algorithm using free I/O mode

The I/O timing of the scheduled design is very different from the behavioral description. Only the input signals **a** and **b** are read in the same clock cycle in the behavioral description and scheduled design. Even the latency of the design is different: 5 clock cycles in the behavioral description; 4 clock cycles in the scheduled design.

But this schedule can provide the designer with important information about the algorithm. In this very simple design, it is easy to see why the output signal **y2** can be calculated earlier than in the behavioral description. But in a complex design, such opportunities for modifying the I/O specification can be much less obvious. The results of scheduling in free mode may prompt the designer to reconsider the original timing specification.

Consider another design that includes a handshaking signal to indicate when the calculation is complete. The behavioral description for such a design in shown in Figure 7-24.

```
main_loop: LOOP
    WAIT UNTIL clk'EVENT AND clk ='1';
    dout <= ( a + b ) * c;
    valid <= '1';
    WAIT UNTIL clk'EVENT AND clk ='1';
    valid <= '0';
END LOOP;
```

Figure 7-24: Simple algorithm with output handshaking signal

Because synthesis can move I/O operations in free I/O mode, I/O operations tend to be placed as early in the schedule as possible. Only data dependencies restrict where the

I/O operations can be scheduled. Figure 7-25 shows this design scheduled in free I/O mode.

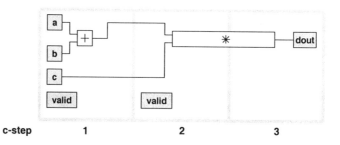

Figure 7-25: Simple algorithm with output handshaking signal scheduled in free I/O mode

Note that the assignment to the signal **valid** is not scheduled in the same c-step as the assignment to the output signal **dout**. This defeats the purpose of the handshaking signal. There is no data dependency that ties these two assignments together. In free I/O mode, these assignments may be scheduled at very different times, as is the case in this schedule.

The schedule in Figure 7-25 illustrates a timing relationship that is maintained when scheduling in free I/O mode. When a behavioral description writes to a signal more than once, the order of the write operations is maintained and the write operations are forced to be scheduled in different c-steps. Thus, the first write to the signal **valid** can be scheduled in the first c-step, because it has no data dependencies to force it later in the schedule. The second write to the signal is placed in the second c-step, because of this rule of free I/O mode scheduling. Of course, scheduling these operations in the first and second c-steps invalidates the desired use of the signal **valid** as a handshaking signal.

This example highlights a limitation of free I/O mode – I/O operations that are used for handshaking can be moved such that their use is invalidated. To address this, behavioral synthesis tools provide commands to establish the timing relationship between I/O (or any other) operations. In the previous example, the designer would need to specify to the synthesis tool that the first write operation to the signal **valid** and the write operation to the signal **dout** must be scheduled in the same c-step.

> While free I/O mode provides the most flexibility for scheduling, it is almost always necessary to define additional tool constraints to force certain I/O operations into particular c-steps or to define a timing relationship between I/O operations.

7.4.2 Testing Designs Scheduled in Free I/O Mode

The previous examples of free I/O scheduling show that synthesized designs will almost always simulate differently than the behavioral descriptions from which they were created. Both the ordering of the I/O operations and the latency of the design can be changed by synthesis.

I/O Scheduling Modes

These differences can greatly complicate testing. One possible solution to this problem is to have two test benches: one for the behavioral description and one for the synthesized design. But a design process that requires different test benches for each stage in the process is likely to be error-prone.

The only other possible solution is to define sufficient I/O timing constraints so that the differences between the I/O timing in the behavioral description and the synthesized design are predictable and thus can be accommodated with a single test bench.

7.4.3 Advantages and Disadvantages of Free I/O Mode

The following lists describe some of the advantages and disadvantages associated with free I/O mode.

Some advantages of free I/O mode are:

- Free I/O mode is the least restrictive scheduling mode and provides for the greatest exploration opportunity.

    The behavioral description does not place any restrictions on the I/O timing or the latency of the synthesized design. Alternate I/O timing protocols can be explored by specifying additional constraints to force certain I/O operations to be scheduled in particular c-steps, or to have a particular scheduling relationship.

- Changes in I/O timing do not require source code modifications.

    Because the timing specified in the behavioral description is not honored when scheduling in free I/O mode, changes in I/O timing are specified entirely with tool-specific commands. A single behavioral description can be used to explore many different I/O timing protocols.

Some disadvantages of free I/O mode are:

- The behavioral description contains no timing information that is used to direct scheduling.

    To specify a particular I/O protocol, timing constraints *must* be specified with tool-specific commands.

    I/O constraints can only be ensured by switching to another scheduling mode (which may require a different coding style) or through the application of scheduling constraints via tool-specific commands.

- The I/O timing of the scheduled design may be very different from the I/O timing of the behavioral description.

    Behavioral synthesis is free to schedule I/O reads and writes in the c-steps that allow for the best use of allocated resources.

    If the I/O timing of the synthesized design is very different from the behavioral description is if difficult, if not impossible, to maintain a single test bench throughout the design process. The design may require 2 test benches (one for the behavioral description and one for synthesized design).

7.5 Summary

This paper discussed the three I/O scheduling modes supported by behavioral synthesis.

Fixed I/O mode is best used when the I/O behavior of a design must be rigidly fixed. A characteristic of such designs is that input and output data are read and produced in a fixed manner at a fixed rate.

Superstate I/O mode is best used when the sequence of I/O operations is important but the exact cycle-to-cycle correspondence between simulation of the behavioral description and the synthesized design is not. This mode is ideal when a handshaking protocol is used to synchronize input and output events, but the number of clock cycles required to perform the calculation is either unknown or may vary depending upon the number and type of available resources.

Free I/O mode is best used when the sequencing of I/O operations is not important or has not yet been defined, or to explore architectural alternatives independent of the limitations that may be imposed by I/O timing requirements. When used in this manner, the results of scheduling in free mode may influence how the I/O timing of a design is ultimately defined. In this mode the designer is exploring architectural alternatives with the fewest constraints.

Each I/O mode has positive and negative characteristics which should be carefully weighed when writing code for behavioral synthesis. Besides the design itself, it is important to consider the supporting environment (such as test benches) when selecting the scheduling mode to be used when synthesizing a design.

Chapter 8

Pipelining

8.1 Types of Pipelining

Behavioral synthesis tools support pipelining techniques to improve the throughput of synthesized designs.

Pipelined components can be used in designs to reduce the latency of a design or to reduce the total area of the components used in the design. Loop pipelining is a feature of behavioral synthesis tools that allows one iteration of a loop to begin before the previous iteration has completed. The structure of a pipelined loop is similar to that of a pipelined component. Loop pipelining can be used to increase the throughput of a loop, but often at the expense of additional component area.

8.2 Pipelined Components

Behavioral synthesis tools support the use of pipelined components in synthesized designs. A pipelined component is a component that can begin a new computation prior to the completion of the current computation. A pipelined component is usually created from a combinational component with similar functionality by dividing the logic of the combinational component into multiple stages and placing registers between the stages to hold intermediate results. The general structure of a pipelined component is shown in Figure 8-1.

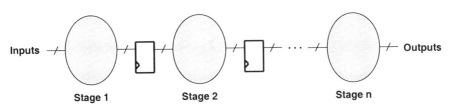

Figure 8-1: The general structure of a pipelined component

> *Pipelined components are characterized by their initialization interval, latency, output delay, and minimum clock period.*

The initialization interval of a pipelined component specifies the rate at which the component can begin processing new input data. Most pipelined components have an initialization interval of 1, which means that the component can begin processing new input data every clock cycle.

The latency of a pipelined component specifies the number of clock cycles it takes to compute output values. This value is one less than the number of stages in the pipelined component.

The output delay of the pipelined component is the combinational delay through the last stage of the pipe. If the pipelined component registers its outputs, the value of the output delay is just the clock-to-output delay of the register.

The longest path between registers within the component determines the minimum clock period. If the clock period were shorter than this value, the component would be unable to correctly capture intermediate values.

Figure 8-2 shows the structure of a pipelined multiplier component and a sample simulation of that component.

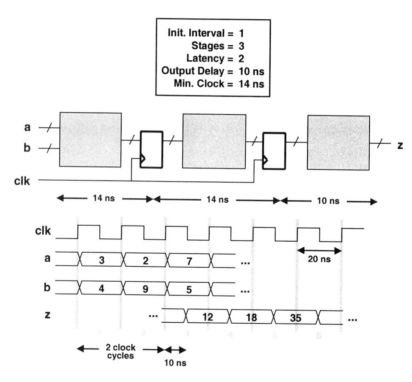

Figure 8-2: The structure and sample simulation for a pipelined multiplier component

The initialization interval of the pipelined multiplier is 1. As shown in the waveforms, new values of the input signals **a** and **b** can be applied every clock cycle. The multiplier has a latency of 2 clock cycles and an output delay of 10 ns. This is illustrated in the

Pipelining

waveforms: the first input values are set to the values 3 and 4 in the first clock cycle, but the first result (12) does not appear at the output of the multiplier until the middle of the third clock cycle.

8.2.1 Design Benefits

Pipelined components can be used in designs to reduce the number of c-steps required to schedule the design or to reduce the total area of the components used in the design. The benefit of pipelined components is illustrated in the following example.

Figure 8-3 shows a portion of code to be scheduled along with a list of components that could be used to implement the design.

```
tmp := ( OTHERS => '0' );
FOR i IN 0 TO 3 LOOP
   tmp := tmp + a(i) * b(i);
END LOOP;
```

Component	Area	Delay
ADD	100	5 ns
FAST_MUL	2000	18 ns
SMALL_MUL	1000	36 ns
PIPE_MUL	1200	Init. Interval = 1 Latency = 1 Delay = 18 ns

Figure 8-3: Code example with components that could be used for implementation

Assume that the calculation must have a latency of no more than 4 clock cycles and that the area of the components used in the design should be minimized.

Note that the multiply operations could all be performed in parallel if sufficient multiplier components were available. There are no data dependencies between the multiply operations.

First consider scheduling the design using adders (the ADD component) and either the fast multiplier component (FAST_MUL) or the small multiplier component (SMALL_MUL). Assume that the clock period is 20 ns.

In order to meet the latency requirement, either 2 fast multipliers or 4 small multipliers would be needed. Schedules for such implementations are shown in Figure 8-4.

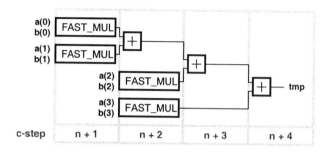

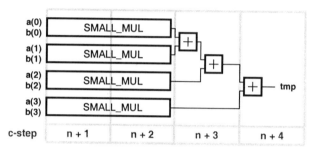

Figure 8-4: Schedules using fast and small multipliers

Both of these implementations require significant area for resources. Now consider using a pipelined multiplier component.

The small combinational multiplier (SMALL_MUL) can be transformed into a 2-stage pipelined multiplier by "splitting" the logic into 2 pieces and inserting registers between the two sections. This is illustrated in Figure 8-5.

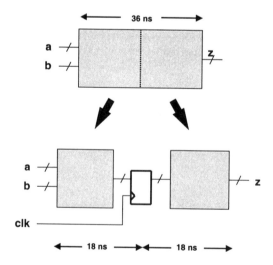

Figure 8-5: Transformation of a combinational multiplier into a 2-stage pipelined multiplier

Pipelining

For ASIC technologies, the area of the pipelined multiplier is marginally larger than that of the small multiplier. The increase in area comes from the area of the registers that must be inserted to store the intermediate values. In many FPGA device architectures there are so many registers that they are considered to be "free", which means that for an FGPA technology, the size of a pipelined component might be the same as its combinational equivalent.

Figure 8-6 shows a schedule that satisfies the latency requirement but uses only 2 pipelined multipliers.

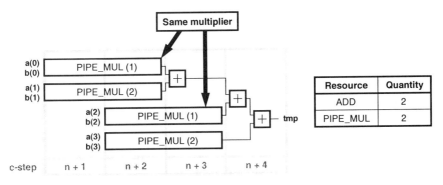

Figure 8-6: Schedules using pipelined multipliers

The first pair of multiplications begin in the first c-step. The second pair of multiplications begin in the second c-step, *before the first pair of multiplications have completed*.

In the second c-step of the Gantt chart, it appears that 4 multipliers are active. But the design only uses 2 multipliers. The overlap illustrated in the second c-step indicates that *both* stages of the multipliers are active in that c-step.

In this example, the pipelined multipliers are effective in reducing the area of the components used to implement the design. This is possible because there are no data dependencies between the multiply operations – it is not necessary to wait for one multiplication to complete before another can begin.

> *In order for pipelined components to be effective in reducing the latency or area of a design, there must be an opportunity for the parallel computation of operations.*

Consider a design with the following statement:

```
y := a * b * c;
```

In this statement, there is no opportunity to perform the multiply operations in parallel – the first multiplication (**a * b**) must complete before the second multiplication can begin. Thus, pipelined multipliers will not help reduce the latency or area of this design.

8.2.2 Design Flow Benefits

Pipelined components can be used in place of multi-cycle components to simplify the back-end RTL design flow. Multi-cycle components complicate the RTL design flow by introducing multi-cycle timing paths between registers that must be specifically identified during gate-level optimization.

Figure 8-7 shows a very simple algorithm, a schedule (that uses a multi-cycle multiplier component), and a hardware implementation of that schedule.

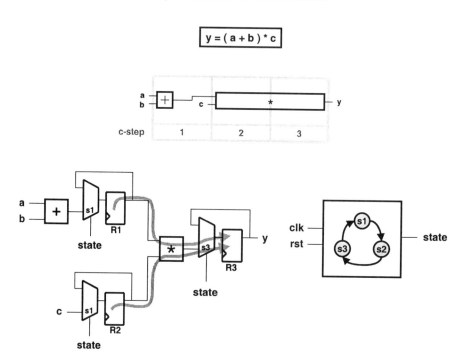

Figure 8-7: Simple design, implemented with a multi-cycle multiplier

The two paths that are highlighted are multi-cycle paths. Since the registers **R1** and **R2** are loaded in state **s1** but the register **R3** is not loaded until state **s3**, the highlighted paths can have a delays as long as 2 clock cycles. Unless such paths are explicitly identified as multi-cycle paths, gate-level optimization tools will assume that the delay between registers that have the same clock must be less than the clock period.

If these paths were not identified as multi-cycle paths, gate-level optimization would attempt to "over-optimize" these paths. The optimization tool could run for a long time, unnecessarily trying to achieve impossible goals. To avoid this potential problem, many RTL designers avoid the use of architectures that would introduce multi-cycle paths into their designs.

When multi-cycle components are used by behavioral synthesis tools, multi-cycle paths can be automatically identified for proper treatment by back-end optimization tools. But if the back-end optimization tool does not support the specification of multi-cycle paths or if a designer has a general practice of never using multi-cycle components, such components can be excluded from possible selection by behavioral synthesis.

The exclusion of multi-cycle components may greatly restrict the architectural alternatives available to the designer. In the example shown in Figure 8-7, the multiply operation would need to be performed by a multiplier whose delay was less than the clock period. The area of such a component may be prohibitively large, or in a given technology, there may not be *any* multiplier components that are fast enough!

A pipelined component can be used to address this issue. Figure 8-8 shows the same design, scheduled using a pipelined multiplier.

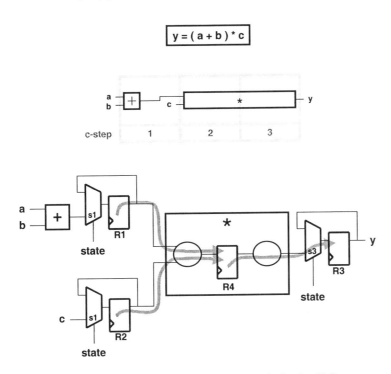

Figure 8-8: Simple design, implemented with a pipelined multiplier

The component labeled **R4** represents the registers that are internal to the pipelined multiplier that are used to hold intermediate values. These registers are loaded every clock cycle.

In this implementation, there are no multi-cycle paths. The paths from the registers **R1** to **R4**, **R2** to **R4**, and **R4** to **R3** are all single-cycle paths.

> *Pipelined components can be used to reduce area without being forced to manage multi-cycle paths in the back-end gate-level optimization flow. Multi-cycle paths are introduced into a design whenever multi-cycle components are used.*

8.3 Loop Pipelining

Loop pipelining is a feature of behavioral synthesis tools that allows one iteration of a loop to begin before the previous iteration has completed. The structure of a pipelined loop is similar to that of a pipelined component. Loop pipelining can be used to increase the throughput of a loop, but often at the expense of additional component area.

To illustrate how a loop is pipelined, a simple non-pipelined loop will be considered. Then, loop pipelining will be applied to increase the throughput of the design.

Figure 8-9 shows a simple algorithm, a timing diagram, and a behavioral description that models the behavior of the algorithm.

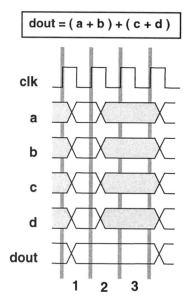

```
main_loop : LOOP

    WAIT UNTIL (clk'EVENT AND clk='1');
    EXIT main_loop WHEN rst='1'

    v_a := a;
    v_b := b;
    v_c := c;
    v_d := d;

    WAIT UNTIL (clk'EVENT AND clk='1');
    EXIT main_loop WHEN rst='1'

    WAIT UNTIL (clk'EVENT AND clk='1');
    EXIT main_loop WHEN rst='1'

    dout <= (v_a + v_b) + (v_c + v_d);

END LOOP main_loop;
```

Figure 8-9: Simple algorithm, timing diagram, and behavioral model

Every 3 clock cycles, the design reads the values of the signals **a**, **b**, **c**, and **d**, and computes the value of **dout**. The timing diagram shows that the computation of **dout** is coincident with the reading of the next values of **a**, **b**, **c**, and **d**.

Assume that the clock period is 10 ns and that the delay through an adder is 6 ns. A possible schedule for the design is shown in Figure 8-10.

Only one adder component is required to implement this design. Even though the schedule contains 3 add *operations*, the adder that is used in the first c-step can be re-used in the second and third c-steps. The behavioral synthesis tool will create the necessary logic and control signals to multiplex data in and out of the adder.

Pipelining

The schedule shown in Figure 8-10 satisfies the timing requirements of the design.

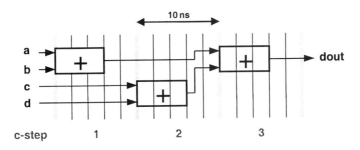

Figure 8-10: Possible schedule for the simple algorithm

8.3.1 Scheduling a Pipelined Loop

Now assume a different timing requirement. Assume that the values of **a**, **b**, **c**, and **d** change *every* clock cycle, but the algorithm can take *up to* 3 clock cycles to compute the value of **dout**. This modified timing diagram is shown in Figure 8-11.

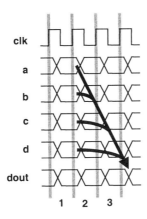

Figure 8-11: Modified timing diagram for the simple algorithm

The throughput of the previous implementation is insufficient to meet the timing requirements of the new specification – a new architecture is required.

One possible approach to this problem is the addition of resources. The addition of resources will often reduce the overall latency of a design by allowing a more parallel implementation.

But for this design, the addition of resources does not help meet this new timing requirement. The entire algorithm would need to be scheduled in a single c-step to satisfy the timing requirement. Even if 3 adders were available, the delay of an adder dictates that only one add operation can be scheduled in a c-step -- the clock period is not long enough to allow multiple add operations to be chained together in a single c-step.

When the addition of resources is not sufficient to produce a design with acceptable throughput, loop pipelining should be considered.

> *Pipelining is a design technique that can be used to increase the throughput of a circuit, usually at the expense of additional hardware. Pipelining can be applied to portions of a circuit that are performed multiple times. In behavioral descriptions, loops represent repeating operations and thus can be pipelined.*

When a loop is scheduled without pipelining, behavioral synthesis tools normally assume that one iteration of a loop must complete before the next iteration of the loop can commence. But if the input data changes more rapidly than the number of c-steps required to schedule the body of the loop, input data can be lost. The modified timing requirements indicate that new input data arrives every clock cycle. But each iteration of the loop requires 3 clock cycles to complete. So, if the schedule in Figure 8-10 is used with the new timing diagram of Figure 8-11, only every *third* set of input data will be processed!

Pipelining the design can solve this problem by allowing subsequent iterations of the loop to begin while current iterations are still processing data.

> *Loop pipelining allows multiple iterations of a loop to operate in parallel.*

Without pipelining, only one c-step in a design is active at a given time. When a loop is pipelined, the c-steps in the Gantt chart become stages of a pipeline that can all be active at the same time. This can be easily visualized as multiple copies of the loop, operating in parallel. Figure 8-12 shows three copies of the loop operating concurrently.

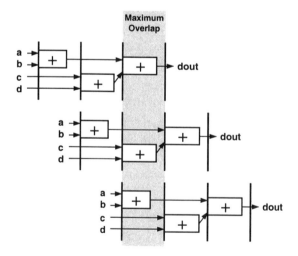

Figure 8-12: Multiple copies of the loop running in parallel

Pipelining

When the first set of input data becomes valid, the first copy of the loop begins processing data. When the second set of input data becomes valid one clock cycle later, the second copy of the loop begins processing that set of data. At the same time the second loop is beginning, the first loop is in the second cycle of the computation. When the third set of input data become valid one clock cycle later, all three copies are operating in parallel.

When the fourth set of input data becomes valid, the first computation is complete. This means that the first copy of the loop can be used again, and so on.

It is not necessary to have hardware for three complete copies of the loop. But any stages that overlap could be active at the same. Thus, there must be sufficient hardware to allow these overlapping stages to perform computations in parallel.

The hardware that is required to implement a pipelined loop can be determined by examining the maximum overlap of multiple copies of the loop running in parallel. This is not a task that must be performed by the user, but does provide a simple way to visualize the results of loop pipelining.

To see the maximum possible overlap, the loop must be copied a sufficient number of times to allow the sequence to repeat. In this example, providing a fourth copy of the loop is not necessary since the first copy can be reused to process the fourth set of incoming data.

The shaded area in Figure 8-12 shows the maximum overlap of these three copies of the loop. This represents the hardware that is required to create a pipelined implementation of the algorithm. The shaded portion of each copy of the loop represents a stage in the pipe.

The hardware that implements a pipelined loop consists of the hardware for each stage of the pipe, separated by registers. The registers pass intermediate results through each stage of the pipe. This structure is shown in Figure 8-13.

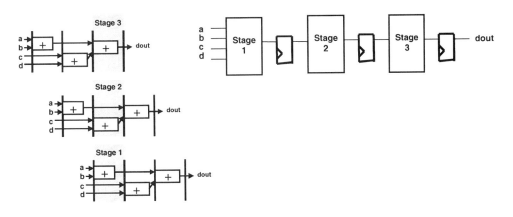

Figure 8-13: Implementation of a pipelined loop

The pipeline requires 3 adders to implement -- one for each stage of the pipe. This represents significantly more resources than the original schedule, which required only a single adder. However, the throughput of the design has been significantly increased. The pipelined implementation can process new data every clock cycle -- the original schedule could only read new input data every 3 clock cycles.

8.3.2 Latency and Initialization Interval

Just like a pipelined component, a pipelined loop has a latency and initialization interval. In the previous example, the initialization interval of the pipelined loop is 1 clock cycle and the latency is 3 clock cycles. This means that new input data can be applied every clock cycle, but it takes 3 clock cycles to produce the result.

Behavioral synthesis tools can pipeline loops with initialization intervals other than 1. Most RTL designers construct pipelined components with an initialization interval of 1, but with the addition of control logic, other intervals can be achieved. In loop pipelining, the initialization interval is driven by the design requirements, resource utilization, and the data dependencies between iterations of the loop.

Figure 8-14 shows a simple algorithm and corresponding schedule.

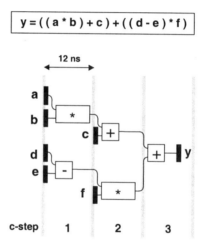

Figure 8-14: Simple algorithm and corresponding schedule

The schedule shows that it takes three c-steps to calculate the value of **y**. The design only requires one multiplier component, since the multiplier used in the first c-step can be reused in the second c-step. Similarly, the add operations in the second and third c-step can be implemented with a single adder component.

Note that the components are not utilized in every c-step. For example, the multiplier is not used in the third c-step, nor is the adder used in the first c-step.

Pipelining

The design characteristics and resource utilization for this schedule are shown in Figure 8-15.

Design Characteristics

Resources	1 multiplier, 1 adder, 1 subtractor
Latency	3 c-steps * 12 ns / c-step = 36 ns
Output Rate	every 36 ns

Resource Utilization

c-step / Resource	1	2	3
Multiply	*	*	
Add		+	+
Subtract	-		

Figure 8-15: Design characteristics and resource utilization

Assume that the throughput of this schedule is not satisfactory. The design could be pipelined to increase the throughput.

Recall that pipelining increases the throughput of a design usually at the expense of additional hardware. But additional resources may not be required if the utilization of resources in the non-pipelined design is low. Notice the empty boxes in the resource utilization table in Figure 8-15. It may be possible to pipeline this design with few or no additional resources.

First consider pipelining this design with an initialization interval of 2. This means that new input data can be supplied every 2 clock cycles (instead of every 3 clock cycles with the non-pipelined design).

The initialization interval determines the number of clock cycles required to complete each stage of the pipeline. When the initialization interval is 1, each stage only requires a single clock cycle to complete. Since this design is being pipelined with an initialization interval of 2, each stage is represented by 2 c-steps. Because the total number of c-steps is 3, the pipelined design will have 2 stages, as shown in Figure 8-16.

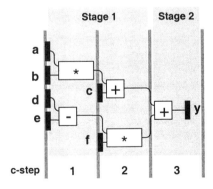

Figure 8-16: Stages of the pipelined design with initialization interval of 2

The loop can be duplicated to show the resources that are required to implement the pipeline. This is shown in Figure 8-17, along with the design characteristics.

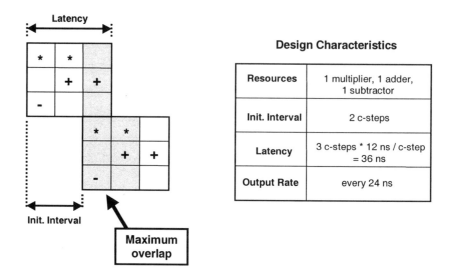

Figure 8-17: Pipelined design with initialization interval of 2

The resources required to implement the pipeline are determined from the maximum overlap between the copies of the loop. Notice that this implementation requires the same resources as the non-pipelined implementation of the design! But the throughput of the design has increased (2 clock cycles vs. 3 clock cycles).

Figure 8-18 shows the utilization of resources for the pipelined design. In this implementation, the components are used almost all of the time. The table shows 2 c-steps, which is the length of each stage of the pipe. Since it is possible for both stages of the pipe to be active at the same time, both stages are displayed in the table.

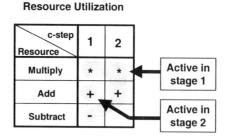

Figure 8-18: Component utilization for pipelined design with initialization interval of 2

Pipelining

> *When component utilization is low, pipelining can increase the throughput of the loop with few or no additional resources.*

If even greater throughput is required, the design could be pipelined with an initialization interval of 1. The loop can be copied to show the resources that are required. This is shown in Figure 8-19.

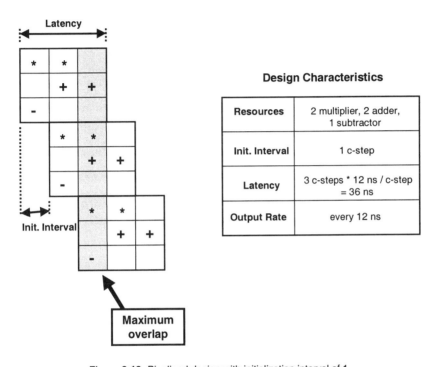

Figure 8-19: Pipelined design with initialization interval of 1

For an initialization interval of 1, additional resources are required. Since there were almost no unutilized resources for an initialization interval of 2, this is not surprising.

8.3.3 NEXT and EXIT Statements in a Pipelined Loop

NEXT and EXIT statements can be used inside a loop which is pipelined. These statements can be used to affect the operation of the stages in the pipe.

Consider a pipeline with an initialization interval of 1 and a latency of 4. Assume the pipeline has a NEXT statement in state 1 and an EXIT statement in stage 2. This is shown in Figure 8-20.

150   **Understanding Behavioral Synthesis**

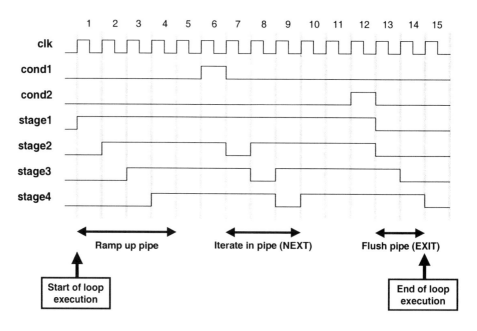

Figure 8-20: Pipelined design with NEXT and EXIT statement

Figure 8-21 shows a simulation of the pipelined loop. The waveform shows how the stages in the pipelined loop are activated or deactivated, based on the signals that control the NEXT and EXIT statements.

Figure 8-21: Simulation of a pipelined design with NEXT and EXIT statement

Pipelining

During the first 4 clock cycles the loop is ramped up. Each stage of the pipe is activated as the first set of data begins flowing through the pipe.

In the sixth clock cycle, the signal **cond1** becomes active, which causes the NEXT statement to be executed. This results in the subsequent stages of the pipe to be deactivated for that iteration of the loop. Note that stage 1 begins processing new data at the next clock cycle.

In the twelfth clock cycle, the signal **cond2** becomes active, which causes the EXIT statement to be executed. This results in the subsequent stages of the pipe to be deactivated. Note that the pipeline continues processing data for 2 clock cycles after the EXIT statement is executed. These cycles are required to flush the pipeline, allowing the calculations that are in progress to complete.

Unexpected behavior can occur when there is an EXIT statement in a pipelined loop in any stage other than the first stage. In this example, the EXIT statement is in the second stage of the pipelined loop. Since multiple copies of the loop are operating in parallel, the first stage of the *next* iteration of the loop has *already occurred* when the EXIT statement is reached! Any variables that are modified in the first stage of the pipeline will have incorrect values and should not be relied upon.

> To prevent incorrect values from being assigned to variables, EXIT statements should only be placed in the first stage of a pipelined loop.

8.3.4 Dependencies in a Pipelined Loop

Data dependencies may prevent a loop for being pipelined. Consider the design shown in Figure 8-22. The figure shows an algorithm and a schedule using a particular multiplier, adder, and subtractor.

```
tmp := ( OTHERS => '0' );

LOOP
    tmp := c * ( ( a + b ) - tmp );
END LOOP;
```

Figure 8-22: Design with dependency between subsequent loop iterations

The value of the variable **tmp** is modified through each iteration of the loop. The value of **tmp** that is calculated in one iteration is used in the next iteration. This implies a data dependency between subsequent iterations of the loop.

Consider pipelining this design with an initialization interval of 1. The loop can be copied to show the required resources. This is shown in Figure 8-23. Data dependencies have been included to show how the value of **tmp** is passed between subsequent iterations of the loop.

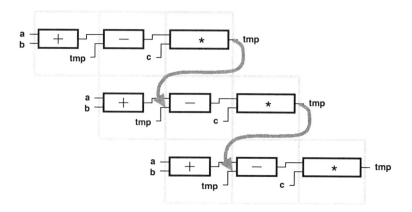

Figure 8-23: Pipelining the design with an initialization interval of 1

The figure shows that this loop cannot be pipelined with an initialization interval of 1. The problem is illustrated by the highlighted data dependencies that flow backward from right to left. This data dependency arrow indicates that the value of **tmp** is needed before it has been calculated!

This design can still be pipelined, but with an initialization interval of 2, as shown in Figure 8-24. Note that the data dependency arrow now points forward.

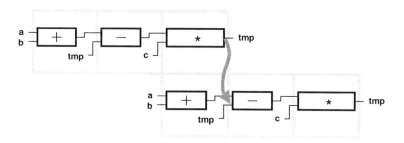

Figure 8-24: Pipelining the design with an initialization interval of 2

8.3.5 Simulating a Pipelined Loop

Pipelining a loop can easily lead to a discrepancy between behavioral simulation and post-synthesis simulation. When a loop is pipelined, multiple iterations of the loop are simultaneously active. This behavior cannot be directly modeled in VHDL.

Pipelining

However, this behavior can be mimicked. To emulate the behavior of a pipelined loop, the following coding style can be used:

- The number of WAIT statements in the loop should correspond to the initialization interval of the loop. This ensures that the loop begins each new calculation at the appropriate time. If a pipelined loop will have an initialization interval of 1, a new calculation should begin every clock cycle.
- Assignments to signals should be delayed by an amount of time that corresponds to the latency of the pipeline. This can be achieved using an AFTER clause.

Figure 8-25 shows a simple design that will be pipelined with an initialization interval of 1 and a latency of 3 clock cycles. The clock period is 10 ns.

```
PROCESS
BEGIN
    dout <= ( OTHERS => '0' );
    main_loop: LOOP
        dout := a * ( b + c ) AFTER 30 ns;
        WAIT UNTIL clk'EVENT AND clk='1';
        EXIT main_loop WHEN  rst = '1';
    END LOOP;
END PROCESS;
```

Figure 8-25: Behavioral description of a design to be pipelined

The body of the loop contains a single WAIT statement, which corresponds to the desired initialization interval of 1. The output is assigned using an AFTER clause. The delay of 30 ns corresponds to a latency of 3 clock cycles.

A simulation of this behavioral description (which will be the same as the pipelined design) is shown in Figure 8-26.

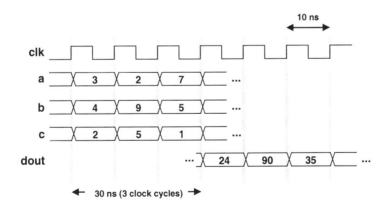

Figure 8-26: Simulation of the behavioral description

The drawback of using this technique is that changes to pipelining (initialization interval or latency) or to clock period (which affects the AFTER clause) must be reflected in the behavioral description.

8.3.6 Restrictions on Pipelined Loops

In order to pipeline a loop, the number of clock cycles required to complete one iteration of the loop must be constant. This is necessary to determine the contents of each stage of a pipeline. This places restrictions of the contents of a loop that is being pipelined.

> *A loop that is being pipelined cannot contain any construct that could require a non-constant number of clock cycles to execute.*

To meet this requirement, behavioral synthesis tools do not allow a loop that is being pipelined to contain any rolled loops or conditional statements with branches of different length. A loop that will be completely unrolled is allowed, as the loop is not present after unrolling.

8.4 Summary

Behavioral synthesis tools support pipelining techniques to improve the throughput of synthesized designs.

Pipelined components can be used in designs to reduce the latency of a design or to reduce the area of the components used in the design. Latency can be reduced because a pipelined component can begin a new computation prior to the completion of the current computation. Pipelined components can also be used in place of multi-cycle components to eliminate multi-cycle paths in the synthesized design.

Loop pipelining is a feature of behavioral synthesis tools that allows a loop to be implemented with the structure of a pipelined component. This technique can be used to increase the throughput of a loop, often at the expense of additional component area. However, if component utilization of a non-pipelined design is low, it may be possible to use pipelining to increase the throughput of the design without additional hardware.

Chapter 9

Memories

9.1 Memories in RTL Design

In VHDL, multi-dimensional arrays provide a powerful and convenient method for modeling the behavior of memories.

But using RTL synthesis, arrays are not easily mapped to memory. If a memory component is required, the designer must either directly instantiate the component in the VHDL description or include a behavioral representation of the memory that can be recognized by the synthesis tool. Either way, the RTL designer must then write code to control all of the signals that affect the operation of the memory: data input and output buses, read / write lines, etc.

When a designer is forced to directly instantiate or model memory, the resulting code is difficult to understand. Describing data using an array is much easier and clearer than directly manipulating memory control lines.

An RTL design must be constructed for a particular memory device. Evaluating the impact of different memories on a particular design is a significant undertaking. If an RTL designer is using an ASIC library that contains both a one-port and a dual-port RAM, he must create two entirely different descriptions. One description instantiates and controls the one-port RAM. The other description instantiates the dual-port RAM and is redesigned to take advantage of the additional port.

Rarely is there time in an RTL design schedule to consider alternative memories and develop multiple descriptions. The designer typically makes a decision about the memories that will be used in a design early in the design process and then "works around" any issues that arise later on.

9.2 Mapping Arrays to Memory

Behavioral synthesis addresses these issues. Behavioral descriptions can manipulate groups of data in an abstract manner using arrays. These arrays can, under the control of the designer, be mapped to memory. Behavioral synthesis tools automatically construct the logic to control the memory, freeing the designer to explore architectures using different memories with different characteristics (e.g. synchronous versus asynchronous, single-port versus dual-port) and to make intelligent decisions about an appropriate implementation for a design.

Figure 9-1 shows VHDL code that declares three variables (**a**, **b**, and **c**) that are 2-dimensional arrays. Each array contains 1024, 8-bit words.

```
SUBTYPE word IS unsigned( 7 DOWNTO 0 );
TYPE array_type IS ARRAY ( integer RANGE <> ) OF word;
VARIABLE a, b, c : array_type( 0 TO 1023 );
```

Figure 9-1: VHDL code that declares a 2-dimensional array variable

The code first declares a SUBTYPE to define the type of each element in the array. The code then defines a type to represent the array itself. Finally, the variables are declared.

When an array is mapped to memory in a behavioral description, accessing the memory is very simple: variable reads and writes correspond to memory reads and writes. For example, the code in Figure 9-2 implies 2 memory reads and 1 memory write. This is certainly easier than writing code to manipulate the memory control signals. Note that a memory read or write is inferred whether the index is constant or non-constant.

```
FOR i IN 0 TO 15 LOOP
    a( i ) := b ( i ) * c ( i ) + 4;
END LOOP;
```

Figure 9-2: Variable reads and writes correspond to memory reads and writes

> *Using HDL arrays as "memory devices" provides memory technology independence as well as easier design development and debugging.*

9.2.1 Mapping Indices to Addresses

When an array is mapped to memory, each element in the array is mapped to a memory location. When only a single variable is being mapped to a memory, the memory location of a particular element corresponds to the index value of that element in the array. This is shown in Figure 9-3.

To map variables to a memory, the designer must specify the type of memory to use and a list of variables that should be mapped to that memory.

Most synthesis tools provide a set of VHDL attributes to specify this information in the behavioral description. To better facilitate the exploration of alternate architectures that utilize *different* memories, some synthesis tools allow this information to be specified or modified from within the synthesis tool itself. Figure 9-4 shows an example of the VHDL attributes that are used to map a variable array to a memory.

Memories

```
SUBTYPE word IS unsigned( 7 DOWNTO
TYPE array_type IS ARRAY ( integer RANGE <> ) OF
VARIABLE mem: array_type( 0 TO 1023
```

	Address
mem(0) →	0
mem(1) →	1
mem(2) →	2
...	...
mem(1022) →	1022
mem(1023) →	1023

Figure 9-3: Each element of the variable is mapped to a corresponding memory location

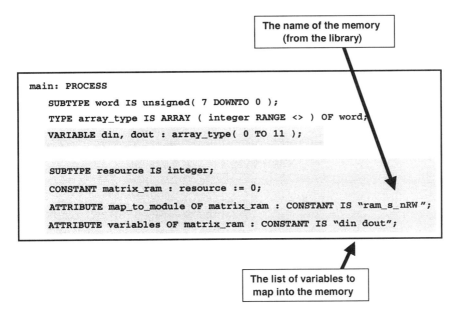

Figure 9-4: VHDL attributes that direct behavioral synthesis to map an array to a memory

The shaded areas in Figure 9-4 highlight the variable declaration and the memory specification.

The memory specification consists of a resource, which indicates that a memory component should be included in the design. A design can declare multiple resources if more than one memory is needed. The resource has an attribute called *map_to_module*, which indicates the type of memory to be used. This name may represent a generic memory model or may be specific to a particular target technology. This design instantiates a memory called **ram_s_nRW**.

The actual behavior of the memory, including the name, phase (active high or low), and use of each signal is described in a library file that is read by the behavioral synthesis tool. This library file also describes the characteristics of the memory, such as number and type (read, write, read / write) of ports and the read and write timing for the memory. This book does not describe the library formats of behavioral synthesis tools.

The resource also has an attribute called *variables,* which indicates the names of the variables to be mapped to this memory. Behavioral synthesis tools allow more than one variable to be mapped to a memory. The example in Figure 9-4 specifies that the variables **din** and **dout** should both be mapped to the same memory.

When more than one variable is being mapped to a memory, some behavioral synthesis tools allow the designer to specify how the indices of the variables should correspond to the addresses in the memory.

One of three different *packing methods* for mapping array indices to memory addresses can be used.

- Explicit

    Using this method, the array indices are taken as explicit addresses. The indices of the arrays being mapped to memory cannot overlap. The variables **din** and **dout** in the description in Figure 9-4 could not be mapped using this method. If the range of the indices for the variable **dout** were changed to "12 TO 23", this method could then be used.

    Explicit packing can lead to portions of a memory being unused if ranges of the indices of the variables being mapped do not adjoin one another. If the range of the indices for the variable **dout** were "13 TO 24", memory address 12 would never be used.

- Packed

    Using this method, the array indices are packed to eliminate gaps in the memory. If the variables **din** and **dout** in the description in Figure 9-4 were mapped using this method, index 0 of the variable **dout** would be mapped to address 12 in the memory. Address 12 is the first available address after **din** has been mapped.

    Additional hardware may be required to transform the array indexing that appears in the behavioral description to the address used by the memory. In this example, the value 12 must be added to every index of **dout**. This transformed address must be calculated prior to the memory access.

- Aligned

    Using this method, the indices for an array are rounded up to the next "power-of-2" boundary. If the variables **din** and **dout** in the description in Figure 9-4 were mapped using this method, index 0 of the variable **dout** would be mapped to address 16 in the memory. Address 16 is the first "power-of-2" address after **din** has been mapped (which ends at address 11).

    Just like packed mode, additional hardware may be required to transform the array indexing that appears in the behavioral description to the address used by the memory. In this example, the value 16 must be added to every index of **dout**. But unlike the example in packed mode, this transformation does not require any additional hardware – it simply requires setting the high-order bit to 1. But the simplification of the indexing comes at a price – a portion of the memory will be unused.

Figure 9-5 shows examples of the three different packing methods.

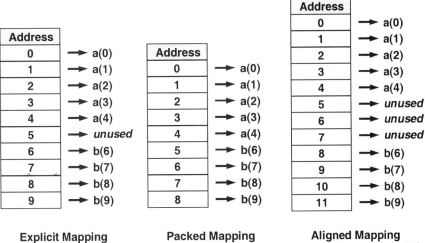

Figure 9-5: Methods for packing multiple arrays into a memory

Variables of differing widths can also be mapped to a single memory. When this is done, certain bits in the memory are never used. Figure 9-6 shows how two variables of differing widths can be mapped to a single memory.

The width of each word in the memory is equal to the width of the widest variable being mapped to that memory. The high-order bits of the words that represent the other variables will be unused. So, this example requires a 9-word by 8-bit memory. Note that packed mapping in used in this example.

Mapping more than one variable to a memory can cause bottlenecks in the design. For example, if the variables **b** and **c** in the example code shown in Figure 9-2 were mapped to the same memory, the 2 memory read operations would need to be scheduled sequentially (assuming a memory with one read port), since both operations require access to the memory. If the variables are mapped to two separate memories, the memory read operations could be performed at the same time.

```
SUBTYPE word1 IS unsigned( 7 DOWNTO 0 );
SUBTYPE word2 IS unsigned( 5 DOWNTO 0 );
TYPE array1_type IS ARRAY ( integer RANGE <> ) OF word;
TYPE array2_type IS ARRAY ( integer RANGE <> ) OF word;
VARIABLE a : array1_type( 0 TO 3 );
VARIABLE b : array2_type( 0 TO 4 );
```

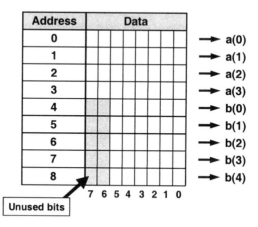

Figure 9-6: Mapping multiple arrays of differing widths into a single memory

9.2.2 Synchronous Memory

Behavioral synthesis tools can automatically map arrays to a wide range of memories with varying characteristics. Asynchronous memories, synchronous memories, memories with one or many ports, and pipelined memories are all examples of memories that can be described for use with behavioral synthesis. This chapter discusses simple synchronous and asynchronous memories to illustrate the general scheduling implications of inferring memories in behavioral descriptions.

Synchronous memories are common in synthesized designs. A synchronous memory reads its input values on clock edges. This integrates very easily with the synchronous nature of synthesized designs. Since synchronous memories contain a bank of input registers, static timing analysis is greatly simplified. From the perspective of static timing analysis, the input to a synchronous memory appears very much like the input to any other register.

The general interface of a synchronous memory is shown in Figure 9-7.

As implied by their name, synchronous memory operations are aligned with a clock. All synchronous memories have a clock signal that captures the value of input signals on a clock edge.

Memories

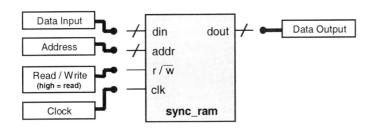

Figure 9-7: General interface of a synchronous memory

Synchronous memories usually have one or more control lines to indicate if data is being read from or written to the memory. This example has a single control line to specify whether data is being read from (active high) or written to (active low) the memory.

Data is written to the memory with the data input bus **din**. Data is read from the memory from the data output bus **dout**. When writing to or reading from the memory, the desired address is specified with the address bus **addr**.

First consider how to write to a synchronous memory. Synchronous memories use input registers to capture input signals on a clock edge. This means that all the signals that are used to write data to the memory must be appropriately set (and stable) some (setup) time prior to the clock edge. At the clock edge, the input values are stored in the input registers of the memory and the data is written to the specified memory address. From the perspective of the external circuit, the write is complete at the clock edge that registers the input data.

Figure 9-8 shows the timing diagram for writing to a synchronous memory.

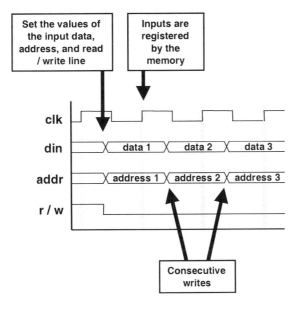

Figure 9-8: Timing diagram for writing to a synchronous memory

Prior to the clock edge, the address, data, and read / write lines are set. The read / write line is set low to indicate a write operation. At the following clock edge, the input data is registered by the memory. Immediately following the clock edge, new data and address values can be set.

The logic that controls the assignment of these signals is automatically generated by behavioral synthesis. The behavioral source code need only contain simple writes to a variable array.

> When mapping an array to memory, the behavioral synthesis tool controls all the signals that are necessary to correctly interface with the memory. This includes read / write lines, address lines, and data lines.

With synchronous memories it is possible to perform consecutive write operations one clock cycle apart. Thus it is possible to schedule write operations in consecutive c-steps (assuming other data dependencies do not cause the operations to be scheduled further apart).

The timing diagram for reading from a synchronous memory is similar to writing to the memory. The input signals must be set prior to the clock edge. The main difference is that the read / write line is high for a memory read. Some delay after the clock edge (as specified by the memory definition) the output data becomes valid and remains stable through the clock cycle. Immediately following the clock edge, new data and address values can be set.

Figure 9-9 shows the timing diagram for reading from a synchronous memory.

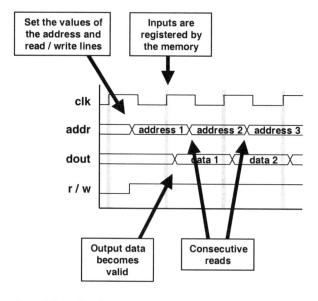

Figure 9-9: Timing diagram for reading from a synchronous memory

With synchronous memories it is possible to perform consecutive read operations one clock cycle apart. Thus it is possible to schedule read operations in consecutive c-steps.

> Both read and write operations for a synchronous memory can be scheduled in one c-step. However, the synchronous nature of the memory requires that these operations be "centered" around a clock edge.

9.2.3 Asynchronous Memory

Behavioral synthesis tools can also map arrays to asynchronous memories. Many ASIC vendor libraries provide both synchronous and asynchronous memories.

Unlike a synchronous memory, an asynchronous memory does not read its input values on clock edges. As the name implies, the interface to such a memory is asynchronous.

The general interface of an asynchronous memory is shown in Figure 9-10.

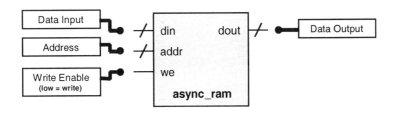

Figure 9-10: General interface of an asynchronous memory

Notice that the interface to the asynchronous memory does not include a clock or a read / write line. Instead, it contains a single write enable signal.

Reading from an asynchronous memory is very straightforward. Essentially, an asynchronous memory is *always* being read. The value of **dout** is dependent only on the address. Any time the address is changed, the value of **dout** will also change.

To a behavioral synthesis tool, reading from an asynchronous memory looks very much like using a combinational component. The desired address is set and some delay later (as specified in the memory definition) the output is valid.

Figure 9-11 shows the timing diagram for reading from an asynchronous memory.

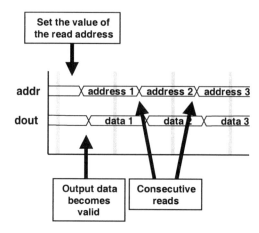

Figure 9-11: Timing diagram for reading from an asynchronous memory

As with synchronous memories, it is possible to perform read operations in consecutive clock cycles. But unlike a synchronous memory, the read operation does not have to be "centered" around a clock edge. When reading from a *synchronous* memory, the address is set in one clock cycle, but the results can not be used until the start of the next clock cycle. This may result in "wasting" a large portion of a clock cycle "waiting" for the next clock edge. This situation does not occur with an *asynchronous* memory. If the address to be read is calculated early in the clock cycle, the contents of the memory could be used in the same clock cycle (if time is available).

Writing to an asynchronous memory is more complicated. For writing, an asynchronous memory behaves very much like a transparent latch. This results in more complicated timing issues.

Assuming an active-low write enable signal, the following timing constraints must be met:

- The address must be stable some time prior to the falling edge of the write enable signal (setup requirement).

- The input data must be stable some time prior to the rising edge of the write enable signal (setup requirement).

- The input data and address must be held stable for some time after the rising edge of the write enable signal (hold requirements).

These timing relationships are shown in Figure 9-12.

These timing relationships have significant impact on the way that write operations are scheduled. To satisfy the setup requirement between the address line and the write enable line (t_{s1}), these signals must be assigned in different c-steps. This is also true of the other setup and hold relationships.

This means that a write operation to an asynchronous memory requires 3 c-steps. In addition, the hold requirements dictate that one memory write must complete before a consecutive write to the memory can begin.

Memories

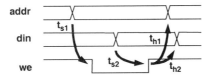

Figure 9-12: Timing relationship for writing to an asynchronous memory

Figure 9-13 shows the timing diagram for writing to an asynchronous memory.

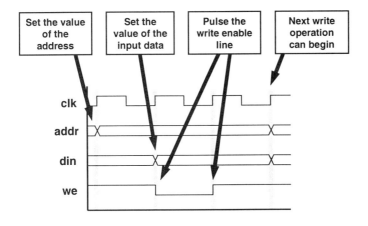

Figure 9-13: Timing diagram for writing to an asynchronous memory

The timing diagram shows how behavioral synthesis will separate the assignment to the memory control signals into different clock cycles. In the first clock cycle of a memory write, the value of the address signal is set; the write enable is inactive. In the second cycle, the value of the input data is set (the data signal only has timing dependencies to the *rising* edge of the write enable signal). Also in the second clock cycle, the write enable is made active. In the third clock cycle, the write enable is once again made inactive. While the write enable signal is latching data (on its rising edge), the address and data signals are held stable. Forcing the signal writes into different clock cycles satisfies the setup and hold requirements of the asynchronous memory.

> *A read operation for an asynchronous memory can be scheduled in a manner similar to a combinational operation and can be performed anywhere within a c-step. However, a write operation requires 3 c-steps, which can result in long schedules.*

9.2.4 Memory Ports

The characteristics of memories can vary widely from one silicon vendor to the next. The most common variation is the type and number of ports.

There are generally three types of memory ports:

- Read ports

    These ports can only read values from the memory. Each read port will have its own address and data out lines.

- Write ports

    These ports can only write values to the memory. Each write port will have its own address, data in, and write enable lines.

- Read / write ports

    These ports can both read values from and write values to the memory, but not at the same time. Each read / write port will have its own address, data in, data out, and read / write lines.

A particular synchronous memory might have 2 read / write ports. Another might have 2 read ports and 1 write port. Different silicon vendors support different variations. Behavioral synthesis tools have the ability to support and take advantage of these different configurations.

Figure 9-14 shows how a simple piece of code can be scheduled. Two schedules are shown for synchronous memories with different port types.

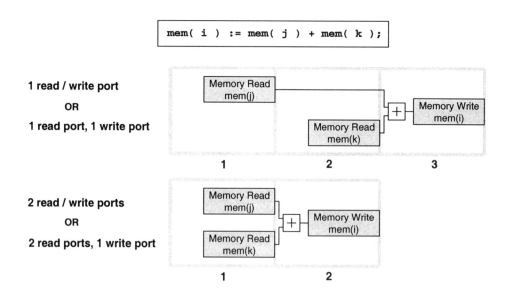

Figure 9-14: Scheduling using different types of synchronous memories

Memories

When the memory has only a single read / write port, the schedule requires 3 c-steps. The statement to be scheduled implies 2 read operations and 1 write operation. Only one of these operations can be done at a time with a read / write port.

Having one read port and one write port does not reduce the schedule. The read operations can only be done one at a time, since there is only one read port. Even though there is a separate write port, the write operation is forced into a later c-step because of data dependencies -- the write cannot be done until the addition is complete (which cannot begin until the memory reads have completed).

However, if the design uses a memory that has the ability to perform the 2 read operations in parallel, the schedule can be reduced to 2 c-steps. For this schedule the read ports are used simultaneously.

> *The number and type of ports on a memory can impact schedule length. Memory exploration is an important aspect of behavioral design.*

9.2.5 Data Dependencies

Behavioral synthesis maintains the order of memory operations. This ensures that correct data is being read.

Consider a behavioral description in which a memory read precedes a memory write.

```
x := mem( i );
mem( j ) := y;
```

Behavioral synthesis places a data dependency between the memory read and the memory write. This ensures that the memory read has completed before the memory write has occurred.

The data dependency is necessary to ensure that the variable **x** is assigned the correct value. Consider a situation in which the values of **i** and **j** are the same. If there were no data dependency between the operations, behavioral synthesis would be allowed to schedule the memory write prior to the memory read. This would result in **x** being assigned the wrong value. In addition, most memories that have multiple ports have a restriction that disallows writing to and reading from the same memory address simultaneously.

Consider a similar situation in which a memory write precedes a memory read.

```
mem( j ) := y;
x := mem( i );
```

The same issue applies. If the two statements were switched in the behavioral code, an incorrect value could be read for **x**.

Consider another code example which has two consecutive memory write operations.

```
mem( i ) := x;
mem( j ) := y;
```

A data dependency is necessary between the two write operations. In the case that the values of **i** and **j** are the same, the second memory write would overwrite the first memory write. If the statements were switched, the contents of the memory would be incorrect.

> *Behavioral synthesis tools adopt a conservative approach to memory inferencing and always create data dependencies between memory reads and memory writes and between consecutive memory writes.*

In some cases this data dependency is unnecessary. Consider another code example.

```
x := mem( i );
mem( i + 1 ) := y;
```

In this code, the index of the memory read and the index for the memory write can never be the same. Thus, the dependency is not necessary.

Behavioral synthesis does not perform the analysis to identify this situation. Although this example appears very obvious, the general problem of determining that two expressions can not have the same value is quite complicated.

However, behavioral synthesis tools provide a mechanism for removing the data dependency between these operations. But the onus is then on the user to ensure that the values of the expression really can never be the same.

In the previous code example, it is obvious that the values can never be the same. But in the first example, where the indices were **i** and **j**, this may be less obvious. If a designer is going to remove the data dependency between two memory operations, it is wise to confirm this decision during behavioral simulation. This can accomplished using an ASSERT statement.

Consider the original code example with the addition of an ASSERT statement.

```
x := mem( i );
mem( j ) := y;
ASSERT( i /= j )
    REPORT "The values of i and j are the same."
    SEVERITY failure;
```

Behavioral synthesis tools ignore ASSERT statements, so the addition of this statement will have no effect on the results of synthesis. However, if the user's assumption that the values of **i** and **j** can never be the same was incorrect, behavioral simulation will report this error.

> *When data dependencies between memory operations are removed, the onus is then on the user to ensure that the values of the expression can never be the same. This can be verified during behavioral simulation by adding an appropriate ASSERT statement.*

Memories

Loop pipelining is another example where data dependencies between operations can cause problems.

Consider a simple loop that shifts the contents of a portion of a memory by 8 locations:

```
FOR i IN 0 TO 7
    mem( i ) := mem( i + 8 );
END LOOP;
```

If the loop is left rolled, the loop requires 24 clock cycles to complete, using the simple synchronous memory previously discussed. Each iteration of the loop requires 3 c-steps to schedule -- the assignment dictates a data dependency between the memory read and the memory write and these operations each require one c-step; the last c-step is used to increment the loop iterator and test the exit condition.

When the loop is left rolled, one iteration of the loop must complete before the next iteration of the loop begins. This is shown in Figure 9-15.

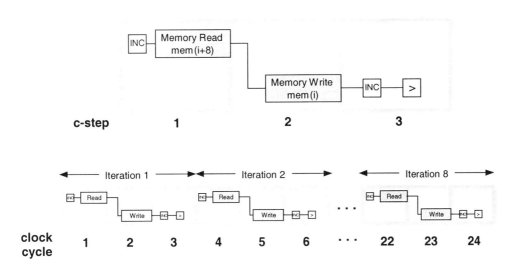

Figure 9-15: Schedule for one iteration of the loop and corresponding clock cycles

This loop can be performed in fewer clock cycles if the loop is pipelined. Figure 9-16 shows the operation of this loop if it is pipelined with an initiation interval of 1, without regard to the data dependencies between the iterations of the loop.

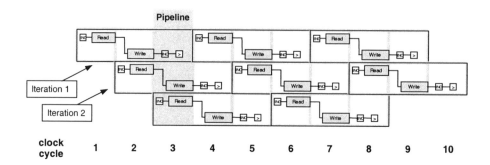

Figure 9-16: Clock cycles required to process the pipelined loop

The figure shows the three stages in the pipeline. Note that there is a memory read in the first stage of the pipe and a memory write in the second stage of the pipe (clock cycle 3). Pipelining this loop can only be accomplished if the memory that has the ability to perform a read operation and a write operation at the same time.

But there is another issue that will prevent this loop from pipelining: the data dependencies between memory operations. Recall that data dependencies are placed between memory reads and writes to ensure that correct data is written to and read from the memory.

Within the loop, there is a data dependency between the memory read and the memory write. This is fine since the data must be read from the memory before it is written to the new location. But there is also a data dependency from the memory write in one iteration of the loop to the memory read operation in the following iteration of the loop. This dependency exists to avoid incorrect behavior. If the index values were the same, the memory would be writing to and reading from the same address simultaneously.

The data dependencies between the memory reads and memory writes are shown in Figure 9-17.

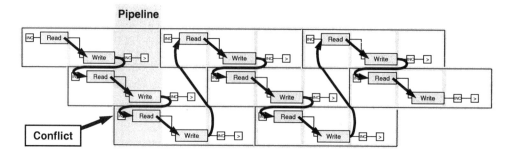

Figure 9-17: Pipelined loop showing data dependencies between memory operations

Memories

The shaded area shows the stages of the pipe. With the data dependencies as shown, the pipeline cannot be created. The memory write and memory read are shown as happening in the *same* c-step, but the data dependency indicates that the memory write must occur *before* the memory read.

In this example, there is no possibility that the index values are the same. This means that it is safe to instruct the behavioral synthesis tool to remove this data dependency. Removing the dependency allows the pipeline to be scheduled.

Consider a modification to the loop that shifts the contents of the memory by only a single location:

```
FOR i IN 0 TO 7
     mem( i ) := mem( i + 1 );
END LOOP;
```

In this case, the overlapping memory write and memory read operations would have the same index. This means that the data dependency cannot be removed.

It is still possible to pipeline the loop, but the initialization interval must be selected such that the data dependency is not violated. This can be accomplished with an initialization interval of 2. The operation of the pipelined design and the data dependencies are shown in Figure 9-18.

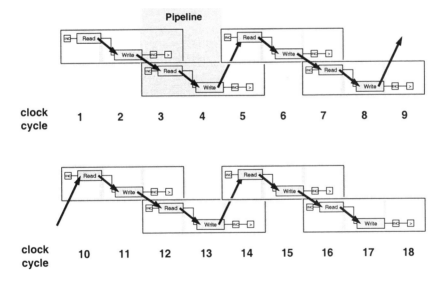

Figure 9-18: Clock cycles required to process the pipelined loop with an initialization interval of 2

The figure shows the two stages of the pipeline. Each stage of the pipe is 2 c-steps, which is the initialization interval. Note that there are no conflicting data dependencies.

Even though this loop requires 18 clock cycles to complete, that is still better than the 24 clock cycles required when the loop is left rolled.

9.3 Summary

Multi-dimensional arrays can be used in behavioral descriptions to manipulate groups of data in a direct, easy-to-understand manner. Behavioral synthesis tools can map these arrays to a variety of types of memories.

Behavioral synthesis tools automatically construct the logic that is necessary to control the memory. This frees the designer to explore architectures using different memories with different characteristics.

Behavioral synthesis tools support many types of memories, including synchronous and asynchronous memories, with varying number and types of ports. Different memory organization strategies can be evaluated when mapping multiple arrays to a single memory.

By evaluating the effect that different types of memories have on schedule length and area usage, the designer can make intelligent decisions about an appropriate memory to include in a design.

Chapter 10

Functions, Procedures and Packages

10.1 Subprograms

In VHDL, the term *subprogram* refers to a function or a procedure. Subprograms can be used in behavioral descriptions to partition the functionality of a design into more manageable, easy-to-understand sections, without hierarchically partitioning the design. This is particularly important for behavioral design, since hierarchical partitioning will likely restrict architectural exploration.

By default, the contents of functions and procedures are *in-lined*, which means they are synthesized as if the contents of the subprogram had been copied into the body of the process where the subprograms are used. Operations inside of subprograms are scheduled along with every other operation in the design.

Consider the two processes shown in Figure 10-1. These processes describe the same functionality, but the process on the left uses a function; the process on the right does not.

```
main: PROCESS
   FUNCTION add( in1, in2: signed(7 DOWNTO 0))
      RETURN signed IS
   BEGIN
      RETURN in1 + in2;
   END add;
BEGIN
   d <= ( OTHERS => '0' );
   main_loop: LOOP
      WAIT UNTIL (clk'EVENT AND clk='1');
      EXIT main_loop WHEN (rst='1');
      d <= add( a, b ) + c;
   END LOOP;
END PROCESS;
```

```
main: PROCESS
BEGIN
   d <= ( OTHERS => '0' );
   main_loop: LOOP
      WAIT UNTIL (clk'EVENT AND clk='1');
      EXIT main_loop WHEN ( rst='1');
      d <= a + b + c;
   END LOOP;
END PROCESS;
```

Figure 10-1: Process with and without a function

The two designs will be synthesized in the same manner -- prior to scheduling, the contents of the function is merged into the body of the process. When a subprogram is in-lined, the resource used to implement the operations in the subprogram can be shared with other operations in the design.

> *When the contents of a function or procedure is in-lined, the operations inside of the subprogram are scheduled along with every other operation in the design.*

10.2 Functions

Some statements can not be used in a function. In particular, VHDL does not allow functions to contain WAIT statements. This affects what can be included in a function when fixed I/O mode is used.

In fixed I/O mode, the statements in the behavioral description that precede a loop or follow a loop must be separated from the loop by WAIT statements. Since WAIT statements cannot appear in functions, a function cannot contain a loop that is not unrolled when scheduling in fixed I/O mode. A loop that will be completely unrolled is allowed, as the loop is not present after unrolling.

By default, the contents of functions and procedures are *in-lined*. Alternately, a function can be mapped to an operator or preserved, as described in the sections that follow.

10.2.1 Mapping to an Operator

Functions can be *mapped* to an operator. Mapping a function to an operator provides a mechanism by which a designer can include (and simulate) pre-constructed components in a design without being forced to actually instantiate those components in the behavioral description.

To understand this feature, it is important to understand the role of operators in the behavioral synthesis process.

The goal of behavioral synthesis is to read a behavioral description and produce a netlist of components. The selection of components to be used in a synthesized design is performed in two steps: translation from HDL code to operators and mapping from operators to components.

In the first step, the HDL code is translated into a Data Flow Graph. The DFG consists of a collection of operators that are connected by dependency arcs. Figure 10-2 shows a simple algorithm and DFG that was first introduced in Chapter 2.

In this design, the DFG contains three operators: ADD, SUB, and MUL.

Behavioral synthesis tools define the set of operators that can be inferred from HDL code.

Particular instances of operators are called *operations*. Each operation has a set of properties that parameterize the operator, such as the width of the inputs and outputs.

Functions, Procedures, and Packages

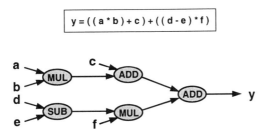

Figure 10-2: Simple algorithm and associated DFG

In the second step, components are selected, based on the *component bindings* that are defined between operators and components. A component binding defines how a component can be used to implement the functionality of the operator.

Component bindings are technology-specific. For example, a particular technology may have a hard macro to perform a 16-bit add. This component is only available when targeting the particular technology. In addition, the component can only be used if the width of the inputs is less than or equal to 16.

Figure 10-3 illustrates the possible bindings of operators to components.

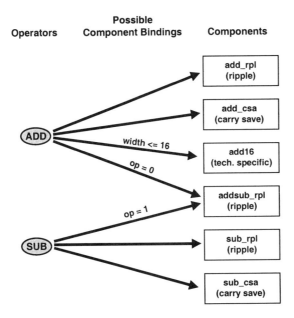

Figure 10-3: Binding of operators to components

Note that there is a choice about how an operation is bound to a component. The figure shows that an ADD operation can be bound to a ripple adder, carry-save adder, or a hard-macro adder. Component bindings can be dependent upon properties of the operator, as is the case for the hard macro.

Component bindings can also specify how the signals of a component should be controlled. For example, both an ADD operation and a SUB operation can be bound to an **addsub** component, but the **op** control line must be set to zero or one, depending upon the operator being bound.

When a function is mapped to an operator, any calls to the function are replaced in the DFG by an operator specified by the designer. The designer can map a function to a "pre-defined" operator (such as ADD) or a new operator can be defined.

If the function is mapped to a predefined operator, then any bindings for that operator can be used. However, if the function is mapped to a new operator, a library with component mappings for that operator must be created. An operator must have at least one component mapping in order to be mapped to hardware.

When a function is mapped to an operator, the contents of the function are ignored during synthesis. The role of the function is to provide a simulation model for verification of the behavioral description. However, the designer must ensure that the function in fact behaves in the same manner as the component. If there are any differences, the pre- and post- synthesis simulation results may differ.

This feature can be used to incorporate intellectual property (IP) blocks into the behavioral synthesis environment, using the following three steps.

- A function must be created that describes the behavior of the block. This function does not have to adhere to the coding guidelines of behavioral synthesis, as the function is not synthesized.

- A new operator must be defined. The function is mapped to this operator.

- One or more component bindings must be defined. If there are multiple implementations of the function, multiple bindings should be specified. This allows the behavioral synthesis tool to select the implementation that best meets the design goals.

> *Function-to-operator mapping can be used to include (and simulate) pre-constructed components in a design without being forced to actually instantiate those components in the behavioral description.*

10.2.2 Preserving a Function

Functions can be *preserved*, if specified by the designer. A preserved function is not in-lined. Instead, the function is scheduled in the same manner as any other operator.

Preserving a function is similar to mapping a function to an operator. When a function is preserved, it is first synthesized as RTL code. The result is a combinational netlist, which represents the component to which the function will be mapped. A new operator is automatically created, and a binding is defined from the new operator to the component. This is just like mapping a function to an operator, but the function actually describes the behavior of the component and the new operator and binding are automatically created.

Functions, Procedures, and Packages

When the design itself is scheduled, the function is treated as a single combinational operation. This means that multiple uses of the function can share the same components, if appropriate. However, the resources used to implement the operations *inside* the function can not be shared with other similar operations in the body of the design.

Figure 10-4 illustrates the scheduling of a preserved function in a design. Assume that the function **f** is a preserved function. The function is called twice in the design, but the schedule uses only one instance of the function, which is shared between the first and second c-steps.

```
main_loop: LOOP
BEGIN
    WAIT UNTIL clk'EVENT and clk = '1';
    EXIT main_loop WHEN rst = '1';
    y <= f( a, b ) + f( c, d );
END LOOP;
```

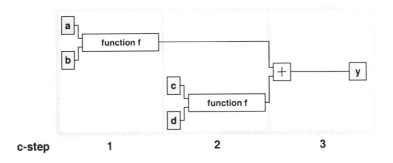

Figure 10-4: Scheduling a preserved function

10.3 Procedures

Procedures are always in-lined. The code inside the procedure is copied into the body of the process wherever the procedure is called. Operations inside of subprograms are scheduled along with every other operation in the design.

Unlike functions, procedures can contain WAIT statements. Thus, virtually any functionality that can be described in the body of a process can be described in a procedure.

The exception to this are EXIT statements that represent reset conditions. These EXIT statements cannot be included in a procedure because VHDL does not allow an EXIT statement to refer to a loop that is outside the scope of the procedure body. When WAIT statements are used inside a procedure, an attribute or synthesis command must be used to declare the reset signal for the design.

Figure 10-5 shows the definition and use of a procedure that includes WAIT statements.

```
PROCESS

   PROCEDURE wait_clock_cycles
      ( CONSTANT n     : IN natural;
        SIGNAL   clk   : IN std_logic;
        CONSTANT phase : IN std_logic ) IS
   BEGIN
      FOR i IN 1 TO n LOOP
         WAIT UNTIL ( clk'EVENT AND clk = phase );
      END LOOP;
   END;

BEGIN

   wait_clock_cycles( 8, clk, '1' );
   y <= ( a + b ) * ( c + d );

END PROCESS;
```

Figure 10-5: Procedure with WAIT statements

10.4 Packages

In VHDL, packages can be used to define type definitions, constant declarations, and subprograms (functions and procedures) outside of a particular design. Packages can be used as part of a design reuse strategy, as the contents of a package can be used in any design.

The items in a package are treated by behavioral synthesis in the same manner as if they were declared directly in the design being synthesized. Whether a function is declared in a package or in an architecture does not affect the results of synthesis.

> *The primary reason for placing definitions or declarations in a package is to allow those items to be reused in another design.*

A subprogram that is highly-specific to a design should probably not be placed in a package. It is unlikely that the function could be reused and would just complicate the package.

Types used on ports must always be declared in external packages. Most common types are declared in readily available packages. For example, the **std_logic** type is defined in the *std_logic_1164* package, which is supported by all popular simulation and synthesis

Functions, Procedures, and Packages

tools. However, if a port must have a user-defined enumerated type, that type would have to be declared in a user-created package.

A package consists of a package declaration and a package body. A package declaration can contain type definitions and constant declarations. A package body is required if subprograms are defined in the package. The interface of the subprograms are defined in the package declaration, but the actual contents of the functions and procedures are defined in the package body.

An example package is shown in Figure 10-6.

```
LIBRARY ieee;
USE ieee.std_logic_1164.ALL;
USE ieee.std_logic_arith.ALL;

PACKAGE useful IS

   PROCEDURE wait_for_high_signal
       ( SIGNAL clock : IN  std_logic;
         SIGNAL sig   : IN  std_logic );

END useful ;

PACKAGE BODY useful IS

   PROCEDURE wait_for_high_signal
       ( SIGNAL clock : IN  std_logic;
         SIGNAL sig   : IN  std_logic ) IS
   BEGIN
      WHILE ( sig /= '1' ) LOOP
         WAIT UNTIL clock'EVENT AND clock = '1';
      END LOOP;
   END wait_for_high_signal;

END useful ;
```

Figure 10-6: Example package

10.5 Summary

Functions, procedures, and packages can all be used in behavioral design.

By default, the contents of functions and procedures are *in-lined*, which means they are synthesized as if the contents of the subprogram had been copied into the body of the design being synthesized. Operations inside of subprograms become indistinguishable from other portions of the design and are scheduled along with every other operation in the design.

A function can be *mapped* to an operator. Mapping a function to an operator provides a mechanism by which the designer can include (and simulate) pre-constructed components in a design without being forced to actually instantiate those components in the behavioral description. When a function is mapped to an operator, the contents of the function are ignored during synthesis. The role of the function is to provide a simulation model for verification of the behavioral description. However, the designer must ensure that the function in fact behaves in the same manner as the component.

A function can be *preserved*, if specified by the designer. When a function is preserved, it is first synthesized as RTL code, independent of the rest of the design. When the design itself is scheduled, the function is treated as a single combinational operation. This means that multiple uses of the function can be shared, if appropriate. However, the operations inside the function can not be shared with other similar operations in the body of the design.

Functions and procedures can be placed in a package, so that they can be used in multiple designs. The items in a package are treated by behavioral synthesis in the same manner as if they were declared directly in the design being synthesized. Whether a function is declared in a package or in an architecture does not affect the results of synthesis. The use of packages is an important element of an overall design reuse strategy.

Chapter 11

Handshaking

11.1 Communication With External Models

An important aspect of any design is how that design communicates with other modules. Previous chapters have shown that many different schedules with different latency can be produced from even a simple piece of code. This complicates communication between multiple designs.

11.1.1 Scheduling Assumptions

Behavioral synthesis tools assume that a process that is being scheduled is part of a synchronous system. Synthesis assumes that the inputs to a process are externally registered. Similarly, synthesis will register the outputs of a scheduled process; output values will only change at clock boundaries. This is shown in Figure 11-1.

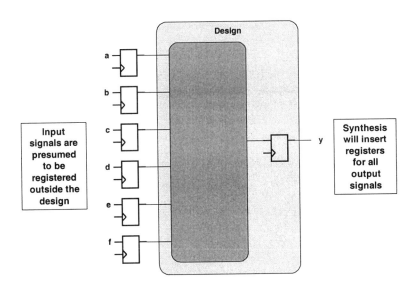

Figure 11-1: I/O assumptions for synchronous systems

These assumptions affect how scheduled designs can communicate with one another. Any communication between designs passes through a register. This implies a one clock cycle delay between the two designs. This is shown in Figure 11-2.

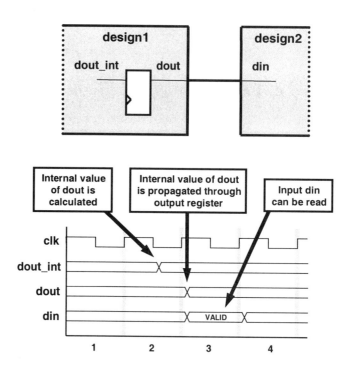

Figure 11-2: Communication between two scheduled designs

Assume the signal **dout** in **design1** is calculated some time in the second clock cycle. Because the output signals of synthesized designs are registered, this value can not be read by the other design until the value has been captured by the register at the following clock edge. Thus, the value of **dout** cannot be read by **design2** until the *third* clock cycle. This delay of one clock cycle must be considered when scheduled designs are passing data.

11.1.2 Synchronizing Communication With Fixed I/O

One way to pass data between models is to synchronize the output writes and input reads to absolute clock cycles. The designs shown in Figure 11-2 could be written such that the signal **dout** is written in c-step 2 and the signal **din** is read in c-step 3. But for this to work, the output write and input read operations would need to be *fixed* to their respective c-steps. This could be accomplished by scheduling both designs in fixed I/O mode (or in another scheduling mode but with additional tool constraints that force the operations into those c-steps).

Handshaking

The behavioral descriptions for two designs that communicate in this manner are shown in Figure 11-3.

```vhdl
LIBRARY ieee;
USE ieee.std_logic_1164.ALL;
USE ieee.std_logic_arith.ALL;

ENTITY design1 IS
    PORT( clk   : IN  std_logic;
          rst   : IN  std_logic;
          a, b  : IN  unsigned( 7 DOWNTO 0 );
          c     : OUT unsigned( 7 DOWNTO 0 ));
END design1;

ARCHITECTURE behavioral OF design1 IS
BEGIN

    p: PROCESS
    BEGIN

        c <= ( OTHERS => '0' );

        main_loop: LOOP

            WAIT UNTIL clk'EVENT AND clk = '1';
            EXIT main_loop WHEN rst = '1';

            c <= a + b;

            WAIT UNTIL clk'EVENT AND clk = '1';
            EXIT main_loop WHEN rst = '1';

        END LOOP;

    END PROCESS;

END behavioral;
```

```vhdl
LIBRARY ieee;
USE ieee.std_logic_1164.ALL;
USE ieee.std_logic_arith.ALL;

ENTITY design2 IS
    PORT( clk   : IN  std_logic;
          rst   : IN  std_logic;
          d, e  : IN  unsigned( 7 DOWNTO 0 );
          f     : OUT unsigned( 7 DOWNTO 0 ));
END design2;

ARCHITECTURE behavioral OF design2 IS
BEGIN

    p: PROCESS
    BEGIN

        f <= ( OTHERS => '0' );

        main_loop: LOOP

            WAIT UNTIL clk'EVENT AND clk = '1';
            EXIT main_loop WHEN rst = '1';

            WAIT UNTIL clk'EVENT AND clk = '1';
            EXIT main_loop WHEN rst = '1';

            f <= d + e;

        END LOOP;

    END PROCESS;

END behavioral;
```

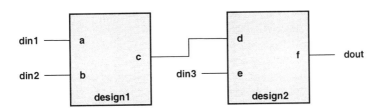

Figure 11-3: Communication between two scheduled designs using synchronized I/O

The design on the left, **design1**, reads the input signals **a** and **b** in the first c-step, calculates the sum of these inputs, and writes the output signal **c**. The design then waits one clock cycle, and repeats.

The design on the right, **design2**, reads the input signals **d** and **e** in the second c-step, calculates the sum of these inputs, and writes the output signal **f**. The design then repeats.

If these designs are both scheduled in fixed I/O mode, the write to the signal **c** in **design1** and the read of the signal **d** in **design2** will be correctly synchronized so the designs can correctly pass data.

For this communication scheme to work, not only must the I/O operations be synchronized, but the latencies of the two designs must also be the same. The designs must have the same latency, so that when the processes in each design repeat, they do so at the same time. This ensures that the data continues to be synchronized for subsequent iterations.

11.2 Handshaking

Unfortunately, this communication scheme is too restrictive for many designs. It forces the designer to fix I/O operations into particular c-steps using fixed I/O mode or additional tool directives. Also, the latency of the design must be constrained. This may overly restrict architectural exploration.

Another approach to this problem involves the introduction of additional signals to indicate when data should be passed between models. This allows the models to independently process data and only synchronize when it is necessary to pass data back and forth. This is called *handshaking*.

Handshaking techniques can be incorporated into the behavioral description so that designs that are synthesized from that description can easily communicate with external models.

11.2.1 Full vs. Partial Handshaking

There are different handshaking protocols that can be used to allow models to successfully communicate with each other.

The degree of handshaking that is required between two modules depends upon whether it is possible for either model to be "busy" when the other model is ready to send data. This chapter discusses two handshaking methods: full handshaking and partial handshaking.

When a model described at the behavioral level is ready to *read* data from an external model, a "ready-for-input" signal could be activated to indicate to the external model that it can begin sending input data. When the external model begins sending data, an "input-valid" signal could be activated so that the model will only read the input data when it is valid.

Similarly, when a model is ready to *write* data to an external model, a "ready-for-output" signal could be examined to check that the external model is ready to receive that data. The model must wait for this signal to become active -- no data can be sent until the external model sets the "ready-for-output" signal active. When this signal is set to active, the model can then write the output data. The model could use an "output-valid" signal so that the external model will read the output data only when it is valid.

Reading and writing data in this manner is called *full handshaking*.

> *When a full handshaking protocol is used, no data is transferred without the acknowledgement of the model to which the data is being sent.*

Handshaking

The general structure of a design that incorporates a full-handshaking protocol is shown in Figure 11-4.

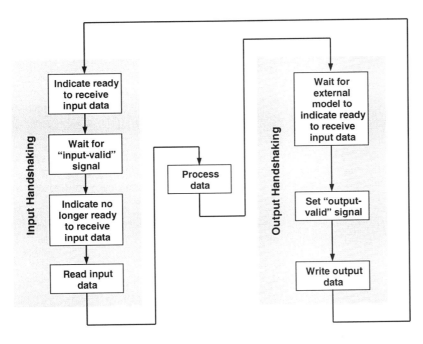

Figure 11-4: The general structure of a design that incorporates a full-handshaking protocol

This approach may be unnecessary for some designs. For example, an external model may process data much more quickly than the model that is sending it data. In this case, there is no need to wait for the external model to be ready to receive data – it is *always* ready to receive data. This eliminates the need for a "ready-for-output" signal.

But the writing of data in one model and the reading of data in another model may still need to be synchronized. Even though the external model is always ready to receive data, it may not know *when* that data will arrive. An "output-valid" signal would still be necessary to indicate that the data being written by one model can be read by another model.

Handshaking that uses an "output-valid" signal, but not a "ready-for-output" signal is called *partial handshaking*.

> *When a partial handshaking protocol is used, data can be transferred without the acknowledgement of the model to which the data is being sent.*

When only partial handshaking is required, some of the blocks in Figure 11-1 can be removed. The general structure of a design that incorporates a partial handshaking protocol is shown in Figure 11-5.

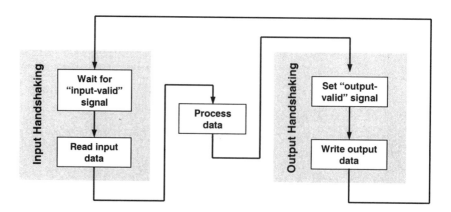

Figure 11-5: The general structure of a design that incorporates a partial-handshaking protocol

The type and amount of handshaking used in a system is of course specific to the flow of data in that system and particularly depends on the relative rates in which models produce and consume data.

Also, the handshaking scheme used to *read* input data may be different from the handshaking scheme used to *write* output data. A design might use full handshaking when reading data and only partial handshaking when writing data. This depends entirely on the characteristics of the other modules with which the design is communicating.

11.2.2 Input Handshaking

Input handshaking is used to correctly pass input data from the outside world to the design. This can be accomplished with two handshaking signals: **ready_for_input** and **input_valid**.

The signal **ready_for_input** is an *output* signal that indicates that the design is ready to accept data.

The signal **input_valid** is an *input* signal that indicates to the design that the input data is valid.

These signals are shown in Figure 11-6.

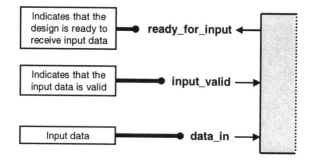

Figure 11-6: Input handshaking signals

Handshaking

Figure 11-7 shows two pieces of code that might be used for input handshaking. These examples were first introduced in Chapter 6. Both pieces of code assume that the signals **ready_for_input** and **input_valid** are active high.

```
main_loop: LOOP
    WAIT UNTIL (clk'EVENT AND clk = '1');
    ready_for_input <= '1';
    WAIT UNTIL (clk'EVENT AND clk = '1');
    WHILE ( input_valid /= '1' ) LOOP
        WAIT UNTIL (clk'EVENT AND clk = '1');
    END LOOP;
    WAIT UNTIL (clk'EVENT AND clk = '1');
    v_a := a;
    v_b := b;
    ready_for_input <= '0';
    WAIT UNTIL (clk'EVENT AND clk = '1');
    ...
END LOOP main_loop;
```

```
main_loop: LOOP
    WAIT UNTIL (clk'EVENT AND clk = '1');
    ready_for_input <= '1';
    WAIT UNTIL (clk'EVENT AND clk = '1');
    IF ( input_valid /= '1' ) THEN
        NEXT main_loop;
    END IF;
    v_a := a;
    v_b := b;
    ready_for_input <= '0';
    WAIT UNTIL (clk'EVENT AND clk = '1');
    ...
END LOOP main_loop;
```

Figure 11-7: Two code examples used to provide input handshaking

The code on the left uses a WHILE loop to wait until the signal **input_valid** is high. The code on the right uses a NEXT statement. The two pieces of may seem very similar but make different assumptions about the timing relationship between the **input_valid** signal and the other input signals.

In the code on the left, the WHILE loop is executed as long as the value of **input_valid** remains at zero. The loop is exited as soon as the value changes to '1'. In this code, the values of **a** and **b** are read in the next clock cycle *after* the signal **input_valid** goes to 1. This is shown in Figure 11-8.

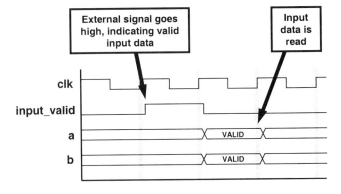

Figure 11-8: Simulation of input handshaking described using a WAIT statement

This may not represent the relationship between these signals. Very often, an "input-valid" signal is set active at the same time the data becomes valid. If the input signals **a** and **b** were valid *only* in the clock cycle when the **input_valid** signal is '1', the input data would be read a clock cycle too late, as shown in Figure 11-9.

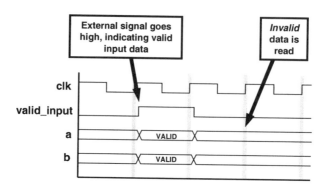

Figure 11-9: Incorrect input handshaking when described using a WAIT statement

It may seem that the way to "fix" the behavior of the code with the WHILE loop is to remove the WAIT statement that is highlighted in Figure 11-4. But the scheduled design will still behave as if the WAIT statement were there – the signals **a** and **b** will be read in the following cycle.

This behavior is a result of the manner in which the state machine is constructed for a WHILE loop. When in a loop, the exit condition is calculated to determine whether to exit the loop or perform another iteration of the loop. If the loop is exited, control is transferred to the state immediately following the loop at the next clock edge. The values of **a** and **b** are read in this state. This means that the earliest time the input signals can be read is one clock cycle *after* the **valid_input** signal goes high. Including the WAIT statement in the behavioral description makes the behavioral design operate in the same manner as the scheduled design.

But changing the input and handshaking signals at the same time is a quite common design practice. This reduces the number of cycles required to read input data by one clock cycle. This scenario can be addressed with the code in Figure 11-7 (on the right) that uses a NEXT statement.

This code uses a NEXT statement to wait until the signal **valid_input** goes to the value '1'. The code waits for a clock cycle and the tests the value of **valid_input**. If the value is 0, the code begins again at the start of the loop, waiting a clock cycle and re-testing the value.

In this code, the input signals **a** and **b** can be read in the same c-step as the signal **valid_input**. The relationship between the behavioral description and wave forms from a sample simulation is shown in Figure 11-10.

Either of these two code examples in Figure 11-7 can be used to implement input handshaking, depending upon the relationship between the handshaking signal and the other inputs to the design.

Handshaking

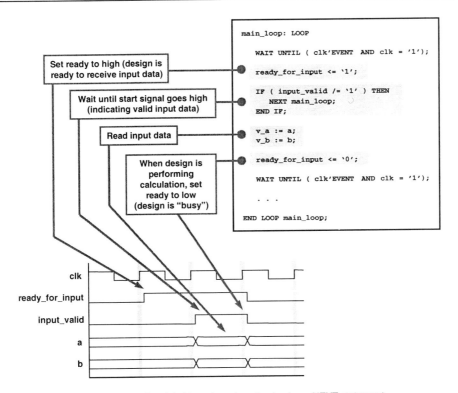

Figure 11-10: Input handshaking when described using a NEXT statement

11.2.3 Output Handshaking

Output handshaking is used to correctly pass output data from the design to the outside world. This can also be coordinated with two handshaking signals: **ready_for_output** and **output_valid**.

The signal **ready_for_output** is an input signal that indicates when the external model is ready to receive input data. This allows the model to "wait" for as long as necessary and prevents the model from writing data that would be "missed" by the external model. This signal is the corollary to the signal **ready_for_input** that is used for input handshaking.

The signal **output_valid** is an output signal that indicates that the output data is valid. These signals are shown in Figure 11-11.

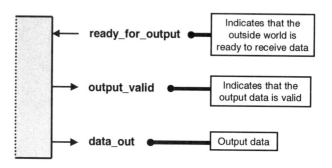

Figure 11-11: Output handshaking signals

The code fragment in Figure 11-12 can be used to control output handshaking. This code fragment assumes that the signals **ready_for_output** and **output_valid** are active-high.

```
main_loop: LOOP

    . . .

    v_data_out := . . .

    WHILE ( ready_for_output = '0' ) LOOP
       WAIT UNTIL ( clk'EVENT AND clk = '1' );
    END LOOP;

    WAIT UNTIL ( clk'EVENT AND clk = '1' );

    data_out <= v_data_out;
    valid <= '1';

    WAIT UNTIL ( clk'EVENT AND clk = '1' );

    valid <= '0';

END LOOP main_loop;
```

Figure 11-12: Code example to provide output handshaking

Handshaking

In this design, the output value is calculated and assigned to an internal variable, **v_data_out**. The code then waits for the external model to be ready to receive the output value. This is accomplished using a WHILE loop that tests the value of **ready_for_output** every clock cycle.

When the **ready_for_output** signal goes active, the model assigns the value to the output signal **data_out** and assigns the value '1' to the signal **output_valid**. The valid signal indicates to the external model that the output data should be read. The **output_valid** signal is held high for one clock cycle, then set back to '0'. This means that the outside world must be "monitoring" the **output_valid** signal so that the output data can be read when appropriate. The input signal **ready_for_output** ensures that this monitoring is taking place.

The relationship between output handshaking signals is shown in Figure 11-13.

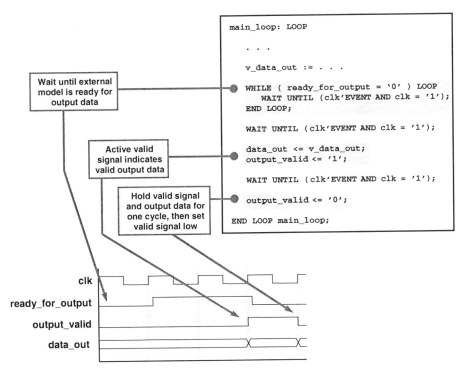

Figure 11-13: Simulation of output handshaking signals

11.2.4 Scheduling Issues

The use of handshaking allows for much greater flexibility when scheduling a design. Because the passing of data between models is synchronized using handshaking signals, the latency of these models can be quite different.

It is necessary, however, for the output-valid handshaking signal and the output data signal to be written in the same c-step. This is necessary to correctly indicate to external models that output data should be read. Fortunately, superstate I/O mode easily addresses this issue.

Consider the behavioral description in Figure 11-14.

```
p: PROCESS
   BEGIN

      dout <= ( OTHERS => '0' );
      valid <= '0';

      main_loop: LOOP

         WAIT UNTIL clk'EVENT AND clk ='1';
         EXIT main_loop WHEN rst ='1';

         dout <= a + b + c + d;
         valid <= '1';

         WAIT UNTIL clk'EVENT AND clk ='1';
         EXIT main_loop WHEN rst ='1';

         valid <= '0';

      END LOOP;

   END PROCESS;
```

Figure 11-14: Behavioral description with output handshaking signal

Assume that this design is scheduled using only a single adder. Because the design contains three add operations, three c-steps are required to calculate the value of **dout** (assuming an add can be performed in one clock cycle).

In fixed I/O mode, this design cannot be scheduled with a single adder. In this mode, the behavioral description specifies that the output must be calculated in a single clock cycle. Since at least three c-steps are required, another scheduling mode must be used.

In superstate I/O mode, the scheduler can "stretch" superstates into multiple c-steps. This would result in the schedule shown in Figure 11-15.

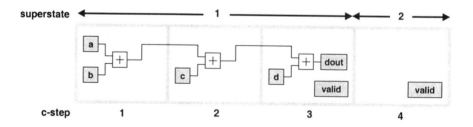

Figure 11-15: Design scheduled in superstate I/O mode

Handshaking

Note that the write to the signal **valid** occurs in the same c-step as the write to the output signal **dout**. Recall that superstate I/O mode forces all write operations that occur within a superstate into the last c-step of the superstate.

> *Superstate I/O mode ensures that output handshaking signals are always correctly synchronized with the associated output data.*

A design that employs a handshaking protocol is most easily scheduled in superstate I/O mode. If the design is scheduled in fixed I/O mode, the appropriate number of clock edges must be added. If the design is scheduled in free I/O mode, the designer must specify additional constraints that ensure that writing output data and writing the corresponding output-valid handshaking signal occur in the same c-step.

In the behavioral description, all of the input signals are read in the first clock cycle. But in the scheduled design, the input signals are read at different times: the signals **a** and **b** are read in the first c-step, **c** is read in the second c-step, and **d** is read in the third c-step.

The input signals are read in different c-steps because they are read in a superstate that had been expanded into multiple c-steps. Recall that superstate I/O mode allows input signals to be read anywhere within a superstate.

If the input signals must all be read in the first c-step, the behavioral description could be modified to separate the input read operations from the calculation of the result, as shown in Figure 11-16.

In this description, a WAIT statement has been added. Variables are introduced to read the input values in the first superstate. These read operations are separated from the calculation and assignment of the result, which occurs in the second superstate. Even if the second superstate is represented by multiple c-steps in the scheduled design, the reading of the input signals is isolated from the assignment of the output by a WAIT statement.

```
p: PROCESS
    VARIABLE v_a : unsigned( 7 DOWNTO 0 );
    VARIABLE v_b : unsigned( 7 DOWNTO 0 );
    VARIABLE v_c : unsigned( 7 DOWNTO 0 );
    VARIABLE v_d : unsigned( 7 DOWNTO 0 );
BEGIN

    dout <= ( OTHERS => '0' );
    valid <= '0';

    main_loop: LOOP

        WAIT UNTIL  clk'EVENT AND clk='1';
        EXIT main_loop WHEN   rst='1';

        v_a := a;
        v_b := b;
        v_c := c;
        v_d := d;

        WAIT UNTIL  clk'EVENT AND clk='1';
        EXIT main_loop WHEN   rst='1';

        dout <= v_a + v_b + v_c + v_d;
        valid <= '1';

        WAIT UNTIL  clk'EVENT AND clk='1';
        EXIT main_loop WHEN   rst='1';

        valid <= '0';

    END LOOP;

END PROCESS;
```

Figure 11-16: Behavioral description modified to isolate input reads from output writes

The schedule that results from using superstate I/O mode with only a single adder resource is shown in Figure 11-17.

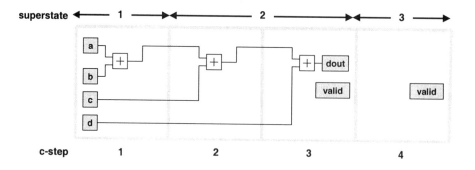

Figure 11-17: Modified design scheduled in superstate I/O mode

Handshaking

11.3 Interprocess Communication

A behavioral description usually contains a single process that reads from and writes to interface signals (ports). But a behavioral description can contain two or more processes. When a description contains multiple processes, data is passed between processes using internal signals.

Behavioral synthesis tools schedule every process in a design separately. Every process can be thought of as having an "interface", which is composed of the set of signals that are read from or written to in the process. These signals can be interface signals (ports) or they can be internal to the design, as shown in Figure 11-18.

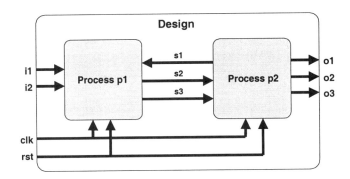

Figure 11-18: The "interface" to a process can include internal and external signals

Because each process is scheduled independently, reading from or assigning to an internal signal is treated by behavioral synthesis in the same manner as an interface signal (port).

> *Communication between processes can be treated in exactly the same manner as communication between a single process and the outside world. The same handshaking techniques that are used to coordinate communication between a model and external blocks can also be used to facilitate communication between two or more processes within the same design.*

11.4 Summary

Developing a communication protocol is an important aspect of any design. Since behavioral synthesis tools register output signals, such signals only change value at clock boundaries. This affects how designs can communicate.

Input read and output write operations in communicating designs can be forced into particular c-steps, ensuring synchronization. This restricts the latency to a set number of c-steps. Unfortunately, this in turn greatly limits architectural exploration.

A simple handshaking protocol can be incorporated into a behavioral description to address the issues of a fixed communication protocol. Handshaking can be used to facilitate predictable communication between designs, but at the same time allow the designer to explore architectures with differing latencies. Full or partial handshaking techniques can be used, depending upon the requirements of the particular design. When scheduling, superstate I/O mode should be used to provide flexibility in scheduling but at the same time ensuring that output signals are appropriately synchronized.

Chapter 12

Reusable Test Benches

12.1 Objectives

Behavioral synthesis allows designers to evaluate alternative hardware architectures with differing clock periods, resources and I/O timing. Based on these user-specified constraints, behavioral synthesis tools schedule operations and produce designs that meet those constraints.

Ideally, a single test bench should be used throughout the design process, since the use of multiple test benches can introduce translation error as well as limit architectural exploration.

This chapter describes a test bench that can be used with a behavioral design, a scheduled design, as well as the resulting optimized gate-level design. This is accomplished via a simple input and output handshaking protocol.

The test bench discussed in this chapter:

- allows easy modification of the clock period.
- supports multiple sets of test data.
- reads input data from a text file and writes output data to a text file for easy insertion into a regression suite.
- employs a simple handshaking protocol to support both pre- and post-scheduled designs.

There is no single style of test bench that is best for all designs. The intent of this chapter is to present an example of a test bench that incorporates a number of different features that should be applicable to many designs.

The design used in this chapter calculates the average of an arbitrary number of signed values.

Data is read serially, one value at a time, via an 8-bit vector, **data_in**. There are two handshaking signals which are used to synchronize input data. The signal **ready** is an output which indicates the design is not currently performing a calculation and can receive new input data. The signal **start** indicates when the first piece of data is valid and that the calculation should commence.

The first piece of data indicates the number of values to be averaged. These values are then read, one per clock cycle.

The input signals and a sample simulation are shown in Figure 12-1.

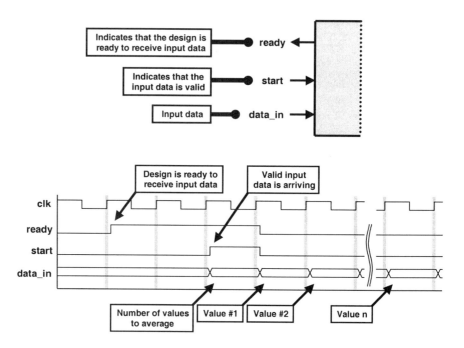

Figure 12-1: Input signals and timing diagram

The design calculates the average value of the input values. When this calculation is complete, the design checks the input signal **output_ready** to see if the external block is ready to receive the result. When this signal is high, the result is assigned to the output port **average**. At the same time, the output signal **valid** is set to high, and held high for one clock cycle. This indicates that the output data is valid.

The output signals and a sample simulation are shown in Figure 12-2.

12.2 I/O Timing

The design uses this full handshaking protocol in order to provide behavioral synthesis with flexibility in scheduling. Although the input data arrives in a set manner, there is no requirement that the average calculation be performed in a particular number of clock cycles. The only requirement is that other blocks that communicate with the design use this handshaking protocol.

With this handshaking protocol it is also possible to create a test bench that can be used through the entire design flow. The test bench will communicate with the design in the same manner as an external block, as described in Chapter 11.

Reusable Test Benches

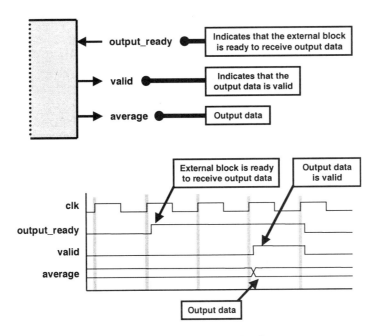

Figure 12-2: Output signals and timing diagram

The test bench that is described in this chapter could be easily adapted to work with a design that is scheduled in fixed I/O mode. When this I/O mode is used, the I/O timing of the synthesized design matches that of the behavioral description. Instead of waiting for a handshaking signal, a test bench for a design scheduled in fixed I/O mode would simply wait the appropriate number of clock cycles before testing the output data.

12.3 Interface Type Considerations

The complete interface to the design is shown in Figure 12-3.

Port	Direction	Description	Type
clk	input	System clock	`std_logic`
rst	input	System reset	`std_logic`
ready	output	Design is ready to receive input data	`std_logic`
start	input	Input data is valid	`std_logic`
data_in	input	Input data	`signed( 7 DOWNTO 0 )`
output_ready	input	External block is ready to receive output data	`std_logic`
valid	output	Output data is valid	`std_logic`
average	output	Output data	`signed( 7 DOWNTO 0 )`

Figure 12-3: Design interface signals

When creating such a design, it may be tempting to declare **data_in** and **average** as type *integer*, rather than *signed*. However, one goal is to create a test bench that could be used *throughout the entire design cycle*. Such a goal has implications on both the test bench *and the design*.

In hardware, there are no abstract types (such as integers), only wires. When a signal is declared to be an integer, the synthesis tool must map that type to a collection of wires. For example, an integer whose values range from 0 to 7 will be mapped into a 3-bit bus.

If a port is declared to be of type integer, the synthesis tool must modify the interface of the design during synthesis. If the interface changes during the design process, then the test bench around the design must also change, to simulate the post-synthesis design.

To create a test bench that does not require any modification, abstract types should be avoided for ports. Thus, the ports **data_in** and **average** do not have integer types. It is still possible to use integer arithmetic, but conversion must be done inside the design. This issue applies to both RTL synthesis as well as behavioral synthesis.

This issue can also be addressed through the use of *wrapper designs*, as discussed in Chapter 5.

The entire example design is shown in Figure 12-4.

Reusable Test Benches

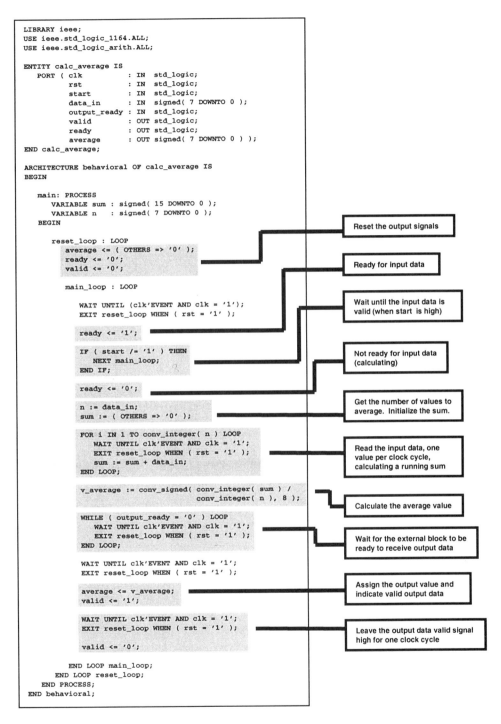

Figure 12-4: Example design

12.4 Test Bench Structure

The purpose of the test bench is to provide input to and analyze the output from a given design. Typically, this input data comprises one or more clock signals, one or more reset signals, and other primary input data. Each of these inputs can be controlled via separate VHDL processes, each running in parallel.

The test bench described in this chapter uses three separate processes to control the interface signals of the design under test. There is a process to control the reset signal, one for the clock signal, and one for the stimulus generation and results verification. Of course, the test bench also instantiates the design to be tested.

For flexibility, the test bench reads input data from an external file and also writes results to a different external file.

The general structure of the test bench is shown in Figure 12-5.

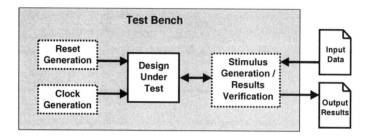

Figure 12-5: General test bench structure

12.4.1 The ENTITY Declaration

Many test benches do not contain SIGNAL or GENERIC declarations. To make the test bench more flexible, the ENTITY declaration of the test bench includes GENERIC declarations to allow modification of certain test bench attributes. These declarations allow modification of the clock period, the name of the file that contains input data, and the name of the file to which output results should be written. A goal is to create a test bench that can be used with different architectures (with potentially different clock periods) and to be able to utilize multiple sets of test data.

Generic values are used because they can be modified without requiring recompilation of the test bench itself. Most VHDL simulators allow modification of generic values on the command line. Thus, different sets of test data require slightly different invocations, but only one test bench is used. This makes the test bench code more maintainable, by eliminating the need for a copy of the test bench for each test to be performed.

The ENTITY declaration for the test bench is shown in Figure 12-6.

Reusable Test Benches

```
ENTITY average_test IS
   GENERIC( clock_delay : time   := 100 ns;
            input_file  : string := "test1_in.txt";
            output_file : string := "test1_out.txt" );
END average_test;
```

Figure 12-6: Entity declaration for the test bench

12.4.2 Instantiating the Design

One of the most basic parts of the test bench is that which instantiates the design under test. This is accomplished via a component declaration and a component instantiation. For each interface signal of the design, there is a corresponding signal which is local to the test bench. These signals are used to communicate between the test bench and the design. The declaration and instantiation of the design is shown in Figure 12-7.

```
ARCHITECTURE behavioral OF average_test IS

   COMPONENT calc_average
      PORT ( clk          : IN  std_logic;
             rst          : IN  std_logic;
             start        : IN  std_logic;
             data_in      : IN  signed( 7 DOWNTO 0 );
             output_ready : IN  std_logic;
             valid        : OUT std_logic;
             ready        : OUT std_logic;
             average      : OUT signed( 7 DOWNTO 0 ) );
   END COMPONENT;

   SIGNAL clk          : std_logic;
   SIGNAL rst          : std_logic;
   SIGNAL start        : std_logic;
   SIGNAL data_in      : signed( 7 DOWNTO 0 );
   SIGNAL output_ready : std_logic;
   SIGNAL valid        : std_logic;
   SIGNAL ready        : std_logic;
   SIGNAL average      : signed( 7 DOWNTO 0 );
   SIGNAL tests_done   : boolean := false;

BEGIN

   -- Instantiate device-under-test.
   dut: calc_average
      PORT MAP( clk          => clk,
                rst          => rst,
                start        => start,
                data_in      => data_in,
                output_ready => output_ready,
                valid        => valid,
                ready        => ready,
                average      => average );
```

Figure 12-7: Declaration and instantiation of the design under test (DUT)

In the code in Figure 12-7, an additional signal has been declared, called **tests_done**. This signal is used to indicate when all the tests have been performed. This provides a mechanism for disabling the clock signal at the end of the test run. Some VHDL simulators have a command that will advance simulation time until all activity has ceased. If the design is simulated with this command, the designer does not have to worry about the exact amount of time required to complete the simulation.

12.4.3 Clock Generation

Designs produced by behavioral synthesis are driven by a system clock. The clock period can be changed to explore different architectures. Thus, it must be easy to modify the clock that is generated by the test bench. As previously mentioned, the clock period is passed into the test bench via a GENERIC declaration.

The test bench contains a process that produces a clock signal with a 50 / 50 duty cycle, based on the generic value. The clock generation process is shown in Figure 12-8.

```
clock_generation:
    PROCESS
    BEGIN
        -- Generate equal duty-cycle clock until all tests are done.
        -- Start clock in an inactive state.
        WHILE ( tests_done = false ) LOOP
            clk <= '0';
            WAIT FOR ( clock_delay / 2 );
            clk <= '1';
            WAIT FOR ( clock_delay / 2 );
        END LOOP;
        -- Wait forever.
        WAIT;
    END PROCESS clock_generation;
```

Figure 12-8: Clock generation process

This process assumes that the clock is active high. The first assignment to the clock is the value '0'. This avoids a clock edge at the very start of the simulation.

The entire clock generation code is enclosed in a conditional that disables the clock when the signal **tests_done** is true, indicating that all tests have completed.

12.4.4 Reset Generation

Almost all designs produced by behavioral synthesis have a reset signal to reset the controller to its start state. The test bench contains a separate process to control this signal.

At the start of the simulation, the reset signal is made active (high, in this example). The signal is held high for three clock cycles, then set low. The reset signal is deactivated in the middle of the clock cycle, to avoid any possible conflict with the active clock edge. The reset generation process is shown in Figure 12-9.

Reusable Test Benches

```
reset_generation:
    PROCESS
    BEGIN
        -- Reset the circuit for 3 clock cycles.
        rst <= '1';
        WAIT FOR clock_delay * 3;
        rst <= '0';
        -- Wait forever.
        WAIT;
    END PROCESS reset_generation;
```

Figure 12-9: Reset generation process

12.4.5 Input and Output Processes

This test bench uses a single process to read input data from an external file, apply that data to the design under test, and evaluate the results. For some designs, it is more convenient to use two processes to accomplish this task. For example, the test bench for the JPEG case study (discussed in Chapter 14) uses two process, as the generation of stimulus and the processing of results are easily separable.

The test bench was designed to read multiple sets of data. Test data is read from an input text file and is written to an output text file. This provides for flexibility in testing. One set of data could be used to perform a "smoke" test (to validate basic functionality), whereas other sets of data could perform more extensive tests. The results are written to an external text file, which allows for easy post-processing (with tools like awk, sed, grep, etc.). Input and output files can be controlled using the GENERIC values on the interface.

The test bench reads and writes ASCII text files. It is possible to write VHDL test benches that read and write binary data. Reading and writing binary data is not discussed in this chapter. An example of this technique can be found in the JPEG test bench, which uses this approach.

A sample input text file is shown in Figure 12-10.

```
#
# n    1    2    3    4    5    6    7    8    ANSWER
##################################################
  8    2    4    6    8   10   12   14   16    9
  8    1    2    3    4    5    6    7    8    4
  8    0    1    2    3    4    5    6    7    4   # Incorrect, should be 3
  4    1    2    3    4                        2
  5    2    2    3    4    4                   3
  5   -2    2    3   -4   -4                   1   # Incorrect, should be -1
```

Figure 12-10: Sample input data file

Every line in the input file represents either a comment line or a set of data to be applied to the design. Comment lines must begin with a hash character (#).

Each line of data begins with the number of values that are being averaged, followed by those values, and finally an expected result. For ease of use and readability, the values appear in the input file as integers and are converted to *signed* types by the test bench.

The test bench also produces an output file that indicates which tests passed or failed.

A sample output text file in shown in Figure 12-11.

```
Test #1, started at 100 ns, passed.
Test #2, started at 750 ns, passed.
Test #3, started at 1400 ns, FAILED - Expected 4, Found 3
Test #4, started at 2050 ns, passed.
Test #5, started at 2500 ns, passed.
Test #6, started at 3000 ns, FAILED - Expected 1, Found -1
```

Figure 12-11: Sample output results file

The test bench uses the text I/O features of VHDL to read input data and write output results. The reading of input data and the analysis of output results are all performed in a single VHDL PROCESS.

The portion of the process that reads the input text file is shown in Figure 12-12.

Reusable Test Benches

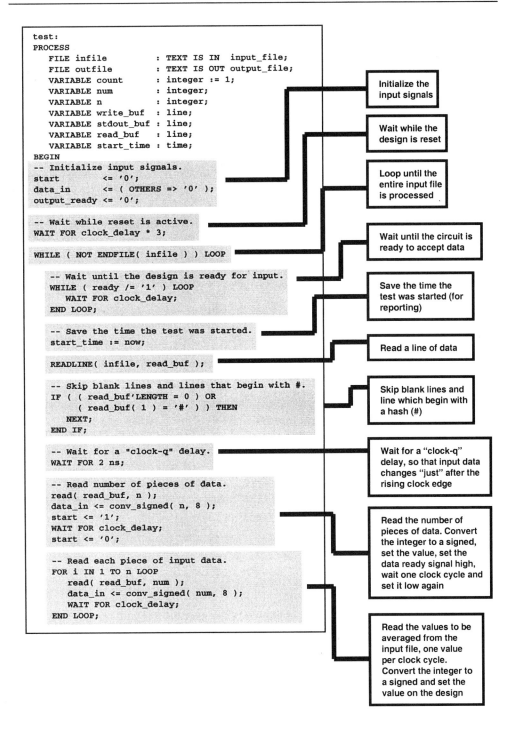

Figure 12-12: VHDL code to read the input text file

This process declares additional variables that are used for reporting purposes. The variable **start_time** stores the time at which the test is started. The variable **count** is incremented after each test, keeping a running count of the number of tests that have been performed.

The text I/O features of VHDL are used to extract data from the input file. The function **readline** reads an entire line from a text file and places the contents of the line in a variable of type *line*. There are a number of **read** functions defined in the *textio* package that extract data from a line. This test bench only uses the function that reads an integer from a line. There are other similar functions that can be used to extract other types of data from a line. Functions that read standard logic types can be found in the package *std_logic_textio*.

The integer values are converted to *signed* using the function **conv_signed**. This function is defined in the *std_logic_arith* package. The second argument to the function specifies the width of the target vector.

Note that the application of the input stimulus is skewed by 2 ns from the clock edge. This small delay mimics the delay of an external register. In general, stimulus should not be modified on an active clock edge. Input signals that change on an active clock edge can introduce iteration loops in zero-delay simulation, and also makes it very difficult to debug a simulation when looking at a waveform display. Debugging is simpler when the inputs do not change at the same time as the clock.

After the input data has been passed to the design, the test bench must now determine if the results are as expected. This is performed in the remaining portion of the process, which is shown in Figure 12-13.

The text I/O package has functions for writing to a file that are similar to the read functions. Writing to a file is done in two steps. First, values are written to a **line**, using the various **write** functions available in the package. Then, the entire line is written to a file using the **writeline** function.

In some of the write statements, an *explicit type cast* is required to remove ambiguity between the various write functions. To a VHDL compiler, a string literal value (such as "abcde") and a *signed* literal value (such as "00011") look the same. If there are two write functions corresponding to these types, the compiler cannot determine which function to use. Note that this problem only arises when literals are used; if a variable was passed to the function, its type would be known and there would not be any ambiguity. An explicit type cast takes the form:

   type_name'(*expression*)

If the test fails (the actual result does not equal the expected result), both the actual and expected are reported in the output file. The conversion function **conv_integer** is used to convert a signed value to an integer.

Reusable Test Benches

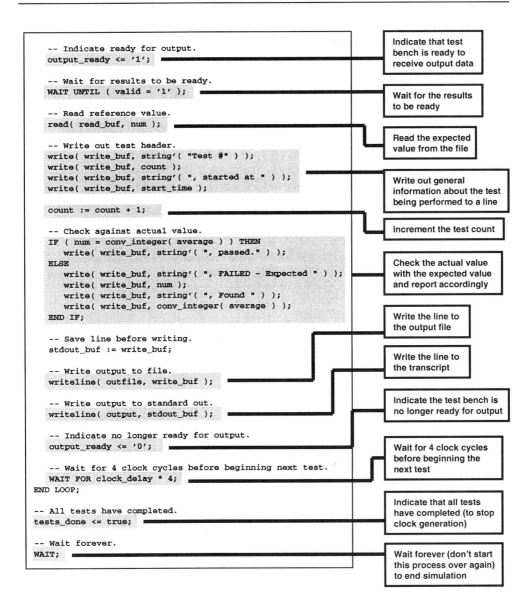

Figure 12-13: VHDL code to process the design output and write to the output file

12.5 Messages

A test bench should produce clear messages, whether a tests passes or fails. Messages can be generated with VHDL using either ASSERT statements or with text I/O. Either method can be used to report the status of the test run.

12.5.1 Assertion Statements

An ASSERT statement is a construct in VHDL that checks if a condition is true or false. If the condition is true, simulation continues. If the condition is false, a message is reported.

Assertions can be assigned a severity. Legal values are: FAILURE, ERROR, WARNING and NOTE. Most simulators treat assertion failures in different ways, based on their severity. For example, an assertion failure with a severity of FAILURE will stop the execution of most simulators by default.

An example ASSERT statement is shown in Figure 12-14.

```
ASSERT ( num = conv_integer( average ) )
   SEVERITY ERROR
   REPORT "Test failed!";
```

Figure 12-14: ASSERT statement to check actual results against expected results

12.5.2 Text I/O

The test bench in this example uses text I/O because it is easier to create more complex formatted output. In addition, messages generated using text I/O can be easily written to external files for post-processing or review at a later time.

Results can be written to the simulator transcript as well as to an external file. This is done in the test bench using the file **output**, which is declared in the VHDL package *standard* (which is always available). The *standard* package declares two special files, **input** and **output**, which correspond to "standard in" and "standard out".

As shown in the test bench, *lines* are written to *files* using functions such as **writeline**. However, care must be taken when writing a line to more than one file. When a line is written to a file, the line is "reset" and its contents are lost. Thus, when writing to two files, the line must be "saved" before it is written to the first file. In the test bench, this is accomplished with an extra variable called **stdout_buf**. The variable **write_buf** is assigned to **stdout_buf** just before the external file is written. Then, the variable **stdout_buf**, which contains the same data, is written to the file **output**.

12.6 Summary

There is no single style of test bench that is best for all designs. This chapter describes a test bench that incorporates a number of different features that should be applicable to many designs.

The test bench allows for easy modification of the clock period, uses a handshaking protocol to apply input stimulus and read output results (thus placing minimal constraints on the scheduling), and is constructed to easily support multiple data sets. This test bench can be used throughout the design process.

A test bench must be tailored to the design to be tested. However, the test bench described in this chapter and the features it contains should provide an adequate starting point for the construction of virtually any test bench.

Chapter 13

Coding For Behavioral Synthesis

13.1 Overview

There are many factors that influence the coding style used for any particular design. Some of these might be:

- Ease-of-development

    Given a particular algorithm described in "c" or a pseudo-language, how quickly can it be translated into VHDL? Some programming constructs, such as pointers, don't readily translate into hardware and are not synthesizable. For these constructs, the designer must develop an alternate representation. Other constructs may be more easily translated. For example, a "struct" in "c" is easily represented as a RECORD in VHDL. Translating portions of an abstract description into VHDL may simplify the development of the behavioral description.

- Understandability

    It is often possible to represent complex data manipulations in a small amount of code. But less code is not always better. What appears obvious to one person may perplex someone else. There is often a trade-off between compactness and understandability.

- Flexibility through synthesis

    In general, a more abstract description will allow a behavioral synthesis tool greater freedom to explore the design space. If a behavioral description is written with a particular hardware implementation in mind, that may largely determine the resultant architecture. Of course, this might not be a bad thing.

- Predictability through synthesis

    This is the corollary to flexibility. If a particular hardware implementation is desired, it may be best to code in a manner that restricts the design space that can be explored.

These are but a few of the many issues that could affect coding style.

> *The desire of most designers is to have set of "rules" that must be followed to process a design successfully and predictably through a behavioral synthesis tool. Unfortunately, it's not that easy.*

This statement is also true for RTL synthesis. Many RTL designers think they can enumerate a list of rules that must be followed for successful RTL design. But these designers are more likely to produce a list of "suggestions" or "issues to consider" than a list of "hard-and-fast" rules. One major difference between coding for RTL synthesis and coding for behavioral synthesis is that RTL synthesis technology is more mature and better understood, so RTL designers can better predict the gate-level constructs which will result from their input descriptions.

While not defining a set of rules, this chapter does develop a set of guidelines would should allow a designer to successfully evaluate a particular piece of behavioral code and understand how it will be processed by a behavioral synthesis tool. This should help a designer approach the comfort level that they may already have with RTL coding guidelines.

The guidelines in this chapter are largely a summation of issues that have been discussed in previous chapters. Where appropriate, the section is referenced.

13.2 Entities, Architectures, and Processes

The ENTITY declaration specifies the input, output, and bi-directional ports that the design uses to communicate with external models. Every port has an associated mode (direction) and type. Any ports with abstract types (such as an enumeration type or an integer) will be changed to non-abstract types during synthesis.

If the interface to a design is modified by synthesis, the resultant netlist can no longer communicate with the surrounding environment, such as the test bench or another behavioral module in the design.

> *Avoid issues associated with type mismatches by only using types that are not changed by synthesis on ports. These types are std_logic, std_logic_vector, signed, and unsigned. Alternately, a "wrapper" model can be created to allow the synthesized design to communicate with the surrounding behavioral environment.*
>
> (Sections 5.1.1 and 5.1.2)

An ARCHITECTURE describes the functionality of a design and can contain concurrent signal assignment statements, component instantiations, and processes.

Coding For Behavioral Synthesis

> *Because behavioral design involves working at a more abstract level than RTL design, it is better to move component instantiations outside of the behavioral description. Inserting technology cells into a behavioral description is somewhat contradictory to the technology-independent nature of behavioral design.*
>
> (Section 5.2)

Some behavioral synthesis tools view concurrent signal assignment statements as representing "glue" logic and thus will not attempt to schedule the operations that appear in those statements. In this case, the statement is synthesized in the same manner as would be performed by an RTL synthesis tool.

> *Only use concurrent signal assignment statements for connecting signals or small amounts of "glue" logic. Include all complex logic in the processes that will be scheduled for a more understandable description.*
>
> (Section 5.2)

In a behavioral description, a process is used to represent the overall flow of an algorithm. For this reason, a behavioral description may contain only a single process to be scheduled.

> *Only partition a design into multiple processes when capacity or run-time is an issue. When a behavioral description is partitioned into multiple processes, architectural exploration is limited. Every process is scheduled independently so no sharing of resources occurs between processes. Also, the designer must ensure that the processes communicate correctly with one another by fixing I/O operations or by using handshaking methods.*
>
> (Sections 5.3 and 11.3)

13.2.1 Specifying Clock Edges

In a behavioral description, a clock edge is specified using a WAIT statement. A process can contain zero or more clock edges. The general structure of a WAIT statement that specifies a rising-edge clock is:

```
WAIT UNTIL clk'EVENT and clk = '1';
```

To specify a falling edge clock, '1' is replaced with '0'.

Clock edges can also be specified using the functions *rising_edge* and *falling_edge* as defined in the *std_logic_1164* package. These functions check the last value of the clock to ensure that clock edge transitions start at '0' or '1' (for rising and falling edges, respectively). An example of a WAIT statement that uses one of the *rising_edge* function is:

```
WAIT UNTIL rising_edge( clk );
```

> *Clock edge specifications that use the rising_edge and falling_edge functions are better simulation models, because transitions that begin at values other than '0' or '1' (such as 'X' or '-') are not valid clock edges.*
>
> (Section 5.3.1)

> *A process can contain multiple WAIT statements, but the WAIT statements must all represent rising or falling edges of the same clock signal. Behavioral synthesis does not allow a process to contain a mix of rising and falling clock edges.*
>
> (Section 5.3.1)

13.2.2 Resets

An important consideration when designing a circuit is providing a method for bringing the circuit into a known state. This is typically done using a reset signal that is made active for a short period of time when power is applied to the circuit. Resetting a circuit usually involves setting output and internal registers to a known value.

> *Any scheduled design that implies a state machine will contain a reset signal. This is because a reset signal is required to initialize the state machine. The reset signal can also be used to initialize other portions of the circuit, but without a reset, the gate-level implementation of the state machine can not be set to a known state.*
>
> (Section 6.3.1)

A reset signal will be added to the synthesized design even if the behavioral description does not explicitly describe the reset behavior. But if the behavior description does not describe the reset behavior, it cannot be tested during behavioral simulation.

Coding For Behavioral Synthesis

> *The most reliable design process includes simulation of the reset behavior. This requires that the reset behavior be explicitly modeled in the behavioral description.*
>
> (Section 6.3.2)

Two examples of the general structure of a process that describes a design with a reset are shown in Figure 13-1. The process contains a "main" loop that describes the normal (non-reset) behavior of the design.

```
PROCESS

    -- Reset action before the main loop is
    -- performed even if reset condition is
    -- NEVER activated!

    main_loop: LOOP
        -- Normal (non-reset) operation

        WAIT UNTIL (clk'EVENT AND clk = '1');
        EXIT main_loop WHEN ( reset = '1' );

        -- Algorithm goes here

    END LOOP main_loop;

END PROCESS;
```

```
PROCESS

    main_loop: LOOP
        -- Normal (non-reset) operation

        WAIT UNTIL (clk'EVENT AND clk = '1');
        EXIT main_loop WHEN ( reset = '1' );

        -- Algorithm goes here

    END LOOP main_loop;

    -- Reset action after the main loop is
    -- ONLY performed even if reset
    -- condition is activated!

END PROCESS;
```

Figure 13-1: Alternate positions for reset action – before (left) and after (right) the main loop

When a reset occurs, the loop is exited, and the reset action, which is described outside of the main loop, is executed. The figure shows the specification of the reset action in two different locations: prior to the main loop and after the main loop.

It is more intuitive to place reset assignments at the start of the behavioral description. However, the model on the right is a better behavioral model because it *requires* that the reset condition be true in order for the reset action to be performed. This means that a test bench must exercise the reset condition even when simulating the *behavioral* model. This is important since a synthesized design will contain a state machine that *must* be reset. Placing the reset action at the end of a behavioral model helps ensure that a test bench or any other model that is interfacing with the description will perform an appropriate reset.

> *It is best to place the reset action after the main loop (instead of before). Although this is a slightly less intuitive code structure (the reset action is the last thing in the behavioral description), the behavioral description will simulate more like the synthesized design.*
>
> (Section 6.3.2)

In a behavioral description, a *synchronous* reset is defined by including an EXIT statement after *every* clock edge. The WAIT and EXIT statements that define a synchronous reset have the form:

```
WAIT UNTIL rising_edge( clk );
EXIT main_loop WHEN rst = '1';
```

> To define a synchronous reset for a behavioral design, an EXIT statement must follow every WAIT statement. These EXIT statements must all have the same exit condition and should exit an infinite loop that describes the normal (non-reset) behavior of the design. The reset behavior is described outside the main loop.
>
> (Section 6.3.1)

An *asynchronous* reset condition is defined with a modified clock definition as well as an EXIT statement. The WAIT and EXIT statements that define an asynchronous reset have the form:

```
WAIT UNTIL rising_edge( clk ) OR rst = '1';
EXIT main_loop WHEN rst = '1';
```

> To define an asynchronous reset for a behavioral design, the clock definition must be modified to include the reset condition. In addition, an EXIT statement must follow every WAIT statement. The WAIT and EXIT statements must all have the same exit condition and should exit an infinite loop that describes the normal (non-reset) behavior of the design. The reset behavior is described outside the main loop.
>
> (Section 6.3.5)

13.3 Data Types

Since a gate-level netlist consists of gates and wires, the job of any synthesis tool (be it behavioral or RTL) is ultimately to translate all data types to those data types which can represent wires.

When considering hardware design, data types can be grouped into two major categories: non-abstract types that readily translate to hardware (e.g. *std_logic*) and more abstract type which have the appearance of software programming constructs (e.g. records).

Coding For Behavioral Synthesis

In VHDL, the *std_logic* type is the standard representation of a wire. During synthesis, all abstract types are translated into the *std_logic* type or the composite (array) types built from *std_logic* (e.g. *std_logic_vector*, *signed*, *unsigned*).

The data types that are supported by behavioral synthesis tools are very similar to those supported by RTL synthesis tools. This is one area where RTL synthesis already permits a quite high level of abstraction. However, even though a large number of data types are supported by both technologies, the way in which types are simulated and synthesized should be considered when selecting data types to be used in a design.

13.3.1 bit / bit_vector Types

The *bit* type is an abstract type that is sometimes used to represent a wire. Because this type has only two possible values ('0' and '1'), there is no way to represent an unknown (or uninitialized) state nor can high-impedance values be represented. A design that is described using the *bit* type will initialize to all zeros by default (even if no reset sequence has been performed). This is a dangerous assumption for simulation.

> *Because the std_logic and std_logic_vector types can easily be used to represent a wire and offer superior functionality, bit and bit_vector types should be avoided.*
>
> (Section 4.2)

13.3.2 boolean Type

The *boolean* type is an abstract type since it is modified by the synthesis process. Like *bits*, *booleans* have only two possible values. The designer should take care to initialize boolean values before they are used. If such an initialization is absent, it will not be detected during behavioral simulation.

> *The value of having boolean flags in a simulation may outweigh its shortcomings. If used in a design, a boolean type should be thought of as an abstract type and thus should not be used to directly represent wires.*
>
> (Section 4.3)

13.3.3 std_logic / std_logic_vector / signed / unsigned Types

The *std_logic* type is the standard representation of a wire. This type and the composite (array) types built from *std_logic* (e.g. *std_logic_vector, signed, unsigned*), are the only types that are found in designs generated using synthesis.

The IEEE 1164 standard defines the *std_logic* logic type as a 9-value enumeration with the values shown in Figure 13-2. These logic values are sufficient to accurately model a wire for simulation.

Value	Meaning
'U'	Uninitialized
'X'	Forcing Unknown
'0'	Forcing 0
'1'	Forcing 1
'Z'	High Impedance
'W'	Weak Unknown
'L'	Weak 0
'H'	Weak 1
'-'	Don't care

Figure 13-2: The 9 states defined for the std_logic type

The interpretation for synthesis of the values '0' and '1' is quite understandable. The values 'L' and 'H' are treated synonymously as '0' and '1', respectively. 'Z' is the high impedance value for a tri-state signal.

The values 'U', 'X', 'W' and '-' are not as well defined for synthesis but they can play an important role in behavioral design. These values can be used to affect simulation results in a manner that does not interfere with behavioral synthesis. These concepts, though not widely used, are also applicable to RTL synthesis.

Unlike many other languages, VHDL requires that every possible condition be accounted for in a CASE statement. This means that for a CASE statement that has a two-bit selection, the bit pattern "0X" must be accounted for as well as "00", "01", etc. For the *std_logic* type that has 9 possible states, this is a very large number of possibilities for even a two-bit value. Fortunately, VHDL has the keyword OTHERS to account for any possible patterns that have not been explicitly listed. This is shown in Figure 13-3.

```
CASE din IS
   WHEN "00"   => dout <= a;
   WHEN "01"   => dout <= b;
   WHEN "10"   => dout <= c;
   WHEN "11"   => dout <= d;
   WHEN OTHERS => NULL;
END CASE;
```

Figure 13-3: CASE statement that does not account for invalid input values

In this example the value of the signal **dout** does not change when the select expression represents invalid data, such as "0X". Even though in real life, patterns such as "0X" can not actually occur (the 'X' value will either be a zero or a one), it is usually beneficial for a

Coding For Behavioral Synthesis

behavioral model to generate "garbage out" when it encounters "garbage in". This CASE statement can be re-written as shown in Figure 13-4 to propagate an unknown value ('X') to the output when invalid input data is encountered.

```
CASE data_in IS
   WHEN "00"  => dout <= a;
   WHEN "01"  => dout <= b;
   WHEN "10"  => dout <= c;
   WHEN "11"  => dout <= d;
   WHEN OTHERS => dout <= (OTHERS => 'X');
END CASE;
```

Figure 13-4: Modified CASE statement that accounts for invalid input values

Similar functionality can be obtained with an IF statement by adding a final ELSE clause as shown in Figure 13-5.

```
IF (data_in = '0') THEN
   dout <= a;
ELSIF (data_in = '1') THEN
   dout <= b;
ELSE
   dout <= (OTHERS => 'X');
END IF;
```

Figure 13-5: IF statement that accounts for invalid input values

> *It is best to create a behavioral design that models "garbage in, garbage out". Design and testing flaws are more likely to be discovered early in the design process and simulation mismatches after synthesis and optimization will be minimized.*

The assignments of unknown ('X') values in the OTHERS clause of the example in Figure 13-4 and in the ELSE clause in Figure 13-5 do not occur in actual hardware. In a physical device, the only possible values of a net are VCC (which is represented by '1') and ground (which is represented by '0') and perhaps 'Z', if the net represents a bus that at times is undriven. The other values that can be represented by the *std_logic* type are artifacts of digital simulation. The purpose of these other values is to detect when nets have not been explicitly assigned a value of '0' or '1'. In hardware, a net will have a value of '0' or '1' when power is applied. The unknown value indicates that whether the net has a value of '0' or a value of '1' can not be predicted. Nevertheless, it will have one value or the other.

> *Assignments of 'X' values under conditions that can not occur in actual hardware can be used to create better behavioral models, but will not affect the results of behavioral synthesis.*

The propagation of unknown values can also be performed by directly examining the values of input or internal signals, rather than as a part of an IF statement or CASE statement, as shown in the previous examples. This provides a more flexible mechanism for propagating unknown value in a behavioral model, but should not generate any hardware, since unknown values do not actually occur in real hardware. Synthesis tools address this issue by translating any comparison to an unknown value to *false*.

Consider the comparison to unknown in Figure 13-6.

```
IF (data_in = 'X') THEN
   -- For simulation ONLY!
   -- These statements will be
   -- removed during synthesis.
   a    <= (OTHERS => 'X');
   b    <= (OTHERS => 'X');
   dout <= (OTHERS => 'X');
END IF;
```

Figure 13-6: Comparison to an "unknown" value

Since the comparison will be translated into the value *false* during synthesis, the statements inside the IF statement could never be activated. The statements are removed and thus not synthesized.

> *For synthesis, direct comparisons to unknown ('X') are mapped to the value "false". If comparisons to 'X' are needed for simulation, they can be safely added to a behavioral model without affecting synthesis.*

For synthesis, comparisons to "don't care" ('-') are treated in the same manner as comparisons to unknown: the comparison is mapped to the value *false*. This means that no additional logic will be produced by synthesis based on comparisons to '-'. However, a designer may want to avoid using comparisons to "don't care". Such comparisons do not simulate in an intuitive manner and may not address any significant modeling issues.

To understand how a comparison to "don't care" works in VHDL, consider the first IF statement in Figure 13-7.

Coding For Behavioral Synthesis

```
IF (din = '-') THEN
   -- Not an input don't care!
END IF;
```

```
IF (din = '0' OR din = '1') THEN
   -- Correct comparison
END IF;
```

```
CASE din IS
   WHEN "00"    =>
   WHEN "01"    =>
   WHEN "1-"    => -- Not an input don't care!
   WHEN OTHERS  => -- "10" and "11" are
                   -- covered here
END IF;
```

Figure 13-7: Incorrect and correct representations of an "input don't care"

If the value of **din** is '0', the comparison is *false*! If the value of **din** is '1', the comparison is also *false*! The only time the comparison is *true* is when the value of **din** is '-'. So, in VHDL, the "don't care" value '-' doesn't really act like a "don't care" at all. This value is a literal value, just like '0', '1', 'X', or 'Z'. Thus, the two IF statements shown in Figure 13-7 are *not* equivalent.

> *The VHDL language does not have a way to specify an "input don't care". A comparison to '-' is literally a comparison to that value and is **not** a comparison to both '0' and '1'.*

Care should also be taken when using '-' in a CASE statement. Figure 13-7 shows the (probably) incorrect use of the "don't care" value in a CASE statement.

13.3.4 Integer Type

Integers are one of the most commonly used abstract types. Integers can be constrained to a particular set of values or can be left unconstrained. An unconstrained integer will be synthesized as 32-bits. Constrained integers will be mapped to the minimum number of bits required to represent the specified range. Positive ranges will be mapped into unsigned representations; ranges that include negative values will be mapped into a 2's complement signed representation.

> *Always constrain the range of integers to the values that are valid for that signal or variable. This will ensure that no unnecessarily large hardware is generated. VHDL simulators perform range checking which will catch "out-of-range" errors during behavioral simulation.*
>
> (Section 4.5)

13.3.5 Enumerated Type

An enumerated type is an abstract type with a discrete set of values. When an enumerated type is synthesized, a unique bit pattern is assigned to each possible value for the enumerated type. By default, bit patterns are assigned to values using a sequential binary encoding scheme.

> *Enumerated types should always be initialized. If they are not initialized, it is possible to "power up" with a bit pattern that does not correspond to any value of the enumerated type. If signals and variables with enumerated types are not initialized, the synthesized circuit may not work correctly.*
>
> (Section 4.6)

13.3.6 Record Type

Behavioral synthesis tools process record types by splitting them into individual elements. This is the equivalent of the user representing each element of a record as an individual signal or variable. Thus, the record type is merely a convenience to the designer: no additional logic is implied when using a record in the place of multiple signals or variables. If a field within a record is also a record, it likewise is split into individual fields.

> *Since no additional hardware is generated when records are used in place of separate assignments, records can be used freely in behavioral descriptions.*
>
> (Section 4.7)

13.3.7 Array Type

Multi-dimensional arrays can be used in behavioral descriptions that will be synthesized. By default, arrays are not mapped to memory and become banks of registers in the synthesized design. Alternately, behavioral synthesis tools can map 2-dimensional

Coding For Behavioral Synthesis

arrays to memory. Whether or not an array is mapped to a memory is under the direct control of the designer.

> *No additional hardware is generated when multi-dimensional arrays are assigned or referenced with constant index expressions. However, large amounts of hardware can be generated when a multi-dimensional array is assigned or referenced with a non-constant index expression. Such use of arrays should be carefully evaluated.*
>
> (Section 4.8)

The designer must consider the hardware that will be generated when using an array in a design. Large arrays whose elements are referenced or assigned with non-constant values should almost always be mapped to memory.

13.3.8 Types Not Supported for Synthesis

Certain types are not supported for behavioral synthesis. These types are: floating point, access, and physical types. (Section 4.9)

13.4 Coding Style and I/O Scheduling Modes

Coding style is greatly affected by the I/O scheduling mode that will be used for a design. For example, in free I/O mode, behavioral synthesis is free to add or delete clock edges and to move I/O operations. This places very few restrictions on the behavioral description. But in fixed I/O mode, the I/O behavior of the synthesized design must match that of the behavioral description. To meet this requirement, the designer must adhere to a strict set of rules. The rules for superstate I/O mode lie somewhere in-between.

13.5 Fixed I/O Scheduling Mode

When a design is scheduled using fixed I/O mode, the I/O timing of the synthesized design is the same as the I/O timing of the behavioral description. The behavioral code describes not only the functionality of the design but also its timing constraints.

In fixed I/O mode, only the I/O operations (signal reads and writes) are fixed to particular clock cycles. Other operations are not fixed, in order to provide greater flexibility in scheduling.

The following coding style guidelines will simplify coding for fixed I/O mode and will result in an easy-to-understand description:

- Place reset actions at either the start or end of the process. The location of the reset actions was discussed in Section 13.2.2.

- Begin the main loop with a WAIT statement. This reflects the actual behavior of the design coming out of reset. After a reset, no assignments are made until the next clock edge.

- If possible, separate the I/O operations from the algorithm. This can be accomplished by creating variables that correspond to the signals that will be read from and written to in the process. Use WAIT statements to read from the external signals and store the values in the variables in the correct clock cycles.
- Place the calculation of the algorithm in one part of the description to enhance readability. Assign results to internal variables.
- Insert the appropriate number of WAIT statements to match the required latency of the design. Then, assign to the output signals.

Consider the simple algorithm and timing diagram shown in Figure 13-8.

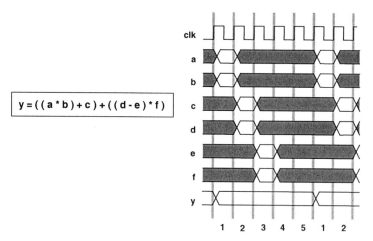

Figure 13-8: Simple algorithm and timing diagram

The application of the suggested coding style to this algorithm is shown in Figure 13-9.

Coding For Behavioral Synthesis

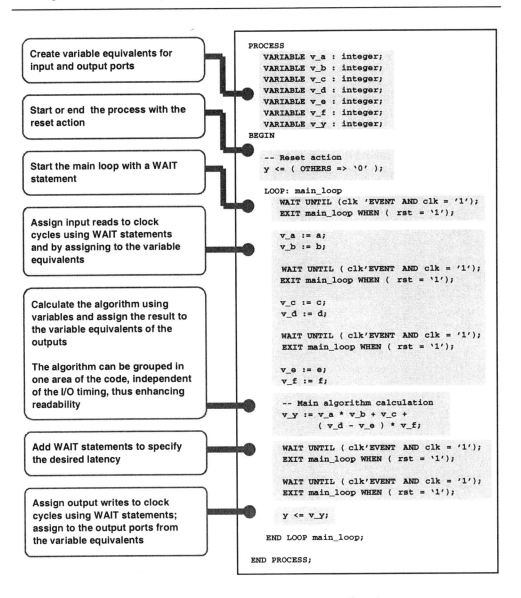

Figure 13-9: Simple algorithm coded for fixed I/O mode

13.5.1 Conditional Statements

Behavioral synthesis tools can schedule conditional statements in two different manners, depending upon the design requirements. The branching in the conditional can either be implemented as multiplexer logic in the data path portion of the design, or as branching in the control portion of the design.

When the branches of a conditional statement contain WAIT statements, the branching is represented in the controller. Each branch can take a different number of clock cycles to execute and is represented by one or more states.

> *When branches of a conditional have unequal length, each branch is represented in the state machine by at least one state. Thus each branch takes at least one clock cycle to execute. This means that in the behavioral description, if any branch contains one or more WAIT statements, then every branch that contains operations must contain at least one WAIT statement.*
>
> (Section 7.2.3)

Examples of conditionals with unequal branch lengths are shown in Figure 13-10.

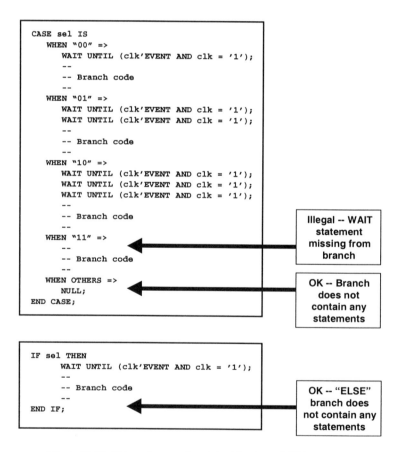

Figure 13-10: All branches must contain at least one WAIT statement

Coding For Behavioral Synthesis

> *In fixed I/O mode, when the branches of a conditional statement contain WAIT statements, it is necessary to isolate the assignments that occur before and after a conditional from the operations inside each branch.*
>
> (Section 7.2.3)

To meet this requirement, the following three conditions must be met when scheduling a conditional with unequal branch lengths in fixed I/O mode:

- Operations that occur *prior* to the start of the conditional must be separated from operations inside each branch by at least one WAIT statement.
- Operations inside each branch of the conditional must be separated from operations that occur *after* the end of the conditional by at least one WAIT statement.
- Operations that occur *prior* to the start of the conditional must be separated from operations that occur *after* the end of the conditional by at least one WAIT statement.

These conditions can be satisfied with the following coding style:

- Begin every branch in the conditional with a WAIT statement.
- Place a WAIT statement immediately following the end of the conditional statement.

These conditions can also be satisfied with an alternate coding style:

- Place a WAIT statement immediately prior to the start of the conditional statement.
- End every branch in the conditional with a WAIT statement.

These coding styles are illustrated in Figures 13-11 and 13-12.

Note that these coding styles are only necessary if the branches have unequal length. Conditionals that do not require branches of unequal length should not contain any WAIT statements.

```
CASE sel IS
   WHEN "00" =>
      WAIT UNTIL ( clk'EVENT AND clk = '1' );
      --
      -- Branch code
      -- (may include additional WAIT statements)
      --
   WHEN "01" =>
      WAIT UNTIL ( clk'EVENT AND clk = '1' );
      --
      -- Branch code
      -- (may include additional WAIT statements)
      --
   WHEN "10" =>
      WAIT UNTIL ( clk'EVENT AND clk = '1' );
      --
      -- Branch code
      -- (may include additional WAIT statements)
      --
   WHEN "11" =>
      WAIT UNTIL ( clk'EVENT AND clk = '1' );
      --
      -- Branch code
      -- (may include additional WAIT statements)
      --
END CASE;
WAIT UNTIL ( clk'EVENT AND clk = '1' );
--
-- Code that follows conditional
--
```

```
IF sel THEN
   WAIT UNTIL ( clk'EVENT AND clk = '1' );
   --
   -- Branch code
   -- (may include additional WAIT statements)
   --
ELSE
   WAIT UNTIL ( clk'EVENT AND clk = '1' );
   --
   -- Branch code
   -- (may include additional WAIT statements)
   --
END IF;
WAIT UNTIL ( clk'EVENT AND clk = '1' );
--
-- Code that follows conditional
--
```

Figure 13-11: Coding style for branches of unequal length for fixed I/O mode

Coding For Behavioral Synthesis

```
--
-- Code that precedes conditional
--
WAIT UNTIL ( clk'EVENT AND clk = '1');
CASE sel IS
   WHEN "00" =>
      --
      -- Branch code
      -- (may include additional WAIT statements)
      --
      WAIT UNTIL ( clk'EVENT AND clk = '1');
   WHEN "01" =>
      -- Branch code
      -- (may include additional WAIT statements)
      --
      WAIT UNTIL ( clk'EVENT AND clk = '1');
   WHEN "10" =>
      --
      -- Branch code
      -- (may include additional WAIT statements)
      --
      WAIT UNTIL ( clk'EVENT AND clk = '1');
   WHEN "11" =>
      --
      -- Branch code
      -- (may include additional WAIT statements)
      --
      WAIT UNTIL ( clk'EVENT AND clk = '1');
END CASE;
```

```
--
-- Code that precedes conditional
--
WAIT UNTIL ( clk'EVENT AND clk = '1');
IF sel THEN
   --
   -- Branch code
   -- (may include additional WAIT statements)
   --
   WAIT UNTIL ( clk'EVENT AND clk = '1');
ELSE
   --
   -- Branch code
   -- (may include additional WAIT statements)
   --
   WAIT UNTIL ( clk'EVENT AND clk = '1');
END IF;
```

Figure 13-12: Alternate coding style for branches of unequal length for fixed I/O mode

13.5.2 Loops

When hardware is generated for a loop, the loop is represented in the controller by one or more states. In the state prior to the start of a loop, a decision is made to either enter the loop or to bypass the loop. In the last state of the loop, a decision is made to either iterate on the loop or to exit the loop.

> *The states that represent a loop are separate from other states in the controller. Thus, in fixed I/O mode, the statements in the behavioral description that precede the loop or follow the loop must be separated from the loop by WAIT statements.*
>
> (Section 7.2.4)

To meet this requirement, the following conditions must be met when scheduling a loop in fixed I/O mode:

- Operations that occur *before* the loop must be separated from operations inside the loop by at least one WAIT statement.

- Operations inside the loop must be separated from operations that occur *after* the loop by at least one WAIT statement.

- When iterating on a loop (because a NEXT statement or the end of the loop has been reached), operations that occur just before iterating on the loop must be separated from operations that occur at the *start* of the loop by at least one WAIT statement.

- Operations that occur *before* the loop must be separated from operations that occur *after* the loop by at least one WAIT statement.

These conditions can be satisfied with the following coding style:

- Start the loop with a WAIT statement.
- Place a WAIT statement immediately following the end of the loop.

These conditions can also be satisfied with an alternate coding style:

- Place a WAIT statement immediately prior to the start of the loop.
- Place a WAIT statement immediately before all NEXT and EXIT statements.
- End the loop with a WAIT statement.

These coding styles are illustrated in Figures 13-13 and 13-14. The first coding style (shown in Figure 13-13) may be preferred, as it requires fewer WAIT statements that could be accidentally omitted.

```
--
-- Code that precedes loop
--
LOOP
   WAIT UNTIL (clk'EVENT AND clk = '1');
   --
   -- Loop code
   -- (may include additional WAIT statements)
   --
   IF next_cond THEN
      --
      -- Conditional code
      -- (may include additional WAIT statements)
      --
      NEXT;
   END IF;
   --
   -- Loop code
   -- (may include additional WAIT statements)
   --
   IF exit_cond THEN
      --
      -- Conditional code
      -- (may include additional WAIT statements)
      --
      EXIT;
   END IF;
   --
   -- Loop code
   -- (may include additional WAIT statements)
   --
END LOOP;
WAIT UNTIL (clk'EVENT AND clk = '1');
--
-- Code that follows loop
--
```

Figure 13-13: Coding style for loops for fixed I/O mode

```
--
-- Code that precedes loop
--
WAIT UNTIL (clk'EVENT AND clk = '1');
LOOP
   --
   -- Loop code
   -- (may include additional WAIT statements)
   --
   IF next_cond THEN
      --
      -- Conditional code
      -- (may include additional WAIT statements)
      --
      WAIT UNTIL (clk'EVENT AND clk = '1');
      NEXT;
   END IF;
   --
   -- Loop code
   -- (may include additional WAIT statements)
   --
   IF exit_cond THEN
      --
      -- Conditional code
      -- (may include additional WAIT statements)
      --
      WAIT UNTIL (clk'EVENT AND clk = '1');
      EXIT;
   END IF;
   --
   -- Loop code
   -- (may include additional WAIT statements)
   --
   WAIT UNTIL (clk'EVENT AND clk = '1');
END LOOP;
--
-- Code that follows loop
--
```

Figure 13-14: Alternate coding style for loops for fixed I/O mode

13.6 Superstate I/O Scheduling Mode

Unlike fixed I/O mode, in which the time between consecutive WAIT statements represents a *single* clock cycle, in superstate mode, consecutive WAIT statements can represent *one or more* clock cycles. In superstate mode, WAIT statements in the behavioral description define the boundaries of a *superstate*.

> *In superstate I/O mode, I/O operations are constrained to the superstate in which they are written and behavioral synthesis is not allowed to move them across superstate boundaries. Synthesis can add as many clock cycles to a superstate as is necessary to schedule the operations in the design, but the <u>ordering</u> of the I/O reads and writes is not changed by synthesis.*
>
> (Section 7.3)

The following coding style guidelines will simplify coding for superstate I/O mode and will result in an easy-to-understand description. These guidelines are very similar to those suggested for fixed I/O mode.

- Place reset actions at either the start or end of the process. The location of the reset actions was discussed in Section 13.2.2.

- Begin the main loop with a WAIT statement. This reflects the actual behavior of the design coming out of reset. After a reset, no assignments are made until the next clock edge.

- If possible, separate the I/O operations from the algorithm. This can be accomplished by creating variables that correspond to the signals that will be read from and written to in the process. Use WAIT statements to read from the external signals and store the values in the variables in the correct clock cycles.

- Place the calculation of the algorithm in one part of the description to enhance readability. Assign results to internal variables.

- Insert the appropriate number of WAIT statements to match the required *minimum* latency (perhaps zero) of the design. Then, assign to the output signals.

Consider the simple algorithm and timing diagram shown in Figure 13-15.

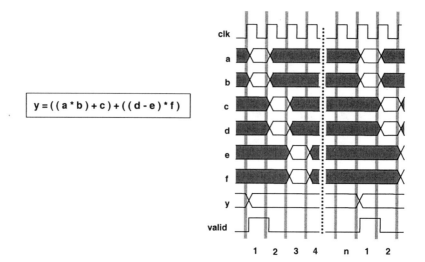

Figure 13-15: Simple algorithm and timing diagram

Just like in the previous example, the timing diagram shows that the input signals **a** and **b** are only valid in the first clock cycle, the input signals **c** and **d** are only valid in the second clock cycle, and the input signals **e** and **f** are only valid in the third cycle. But in this example, the number of cycles required to calculate the result is not specified. The diagram does, however, show that the signal **valid** goes high when the output signal **y** is valid, then goes back low while the next result is being calculated.

The application of the coding style to this algorithm is shown in Figure 13-16.

Coding For Behavioral Synthesis

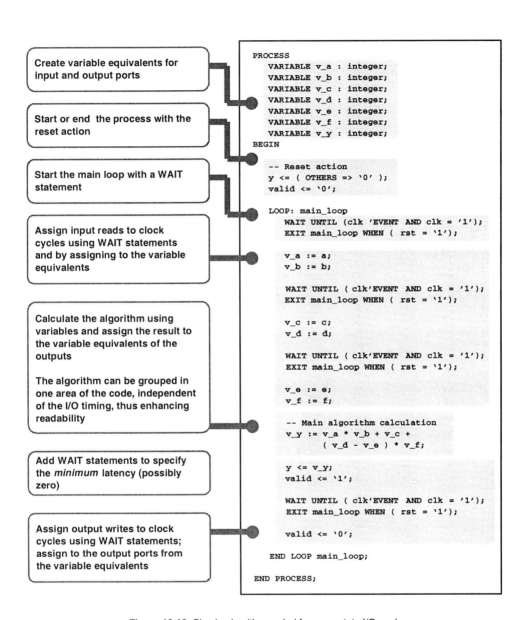

Figure 13-16: Simple algorithm coded for superstate I/O mode

13.6.1 Conditional Statements

Conditional statements (IF and CASE statements) can be scheduled in two different manners in superstate I/O mode.

If there are *no* WAIT statements in any branch of the conditional, then the branching logic will be represented in the data path, as it is with fixed I/O mode when no WAIT statements are used.

But unlike fixed I/O mode, in superstate I/O mode the number of cycles required to process a conditional statement can be extended by the behavioral synthesis tool with the addition of c-steps into a superstate. However, after scheduling, each branch in the conditional will be scheduled in the same number of c-steps.

When a conditional statement contains WAIT statements, each branch is represented in the controller as separate sets of states, similar to fixed I/O mode. For such conditionals, three conditions must be met. These are the same conditions that must be met when scheduling a conditional with unequal branch lengths in fixed I/O mode:

- Operations that occur *prior* to the start of the conditional must be separated from operations inside each branch by at least one WAIT statement.

- Operations inside each branch of the conditional must be separated from operations that occur *after* the end of the conditional by at least one WAIT statement.

- Operations that occur *prior* to the start of the conditional must be separated from operations that occur *after* the end of the conditional by at least one WAIT statement.

These rules also isolate the superstates that precede and follow the conditional from the conditional itself. When a branch of a conditional is scheduled as a superstate, the superstate cannot extend beyond the boundaries of the conditional statement.

These conditions can be satisfied with the following coding style, which is the same as that for fixed I/O mode:

- Begin every branch in the conditional with a WAIT statement.

- Place a WAIT statement immediately following the end of the conditional statement.

These conditions can also be satisfied with an alternate coding style:

- Place a WAIT statement immediately prior to the start of the conditional statement.

- End every branch in the conditional with a WAIT statement.

These coding styles were illustrated in Figures 13-11 and 13-12 in the discussion of fixed I/O mode.

Coding For Behavioral Synthesis

Similar to fixed I/O mode, these coding styles are only necessary if the branches have unequal length. Conditionals that do not require branches of unequal length should not contain any WAIT statements.

13.6.2 Loops

Previous discussions have shown how loops are represented by one or more states in the controller. The operation of a loop can be thought of as being "isolated" from the operation of the rest of the design.

This has implications for scheduling. Assignments to signals (signal write operations) that occur before a loop in a behavioral description are scheduled prior to the start of the loop. Similarly, assignments to signals that occur inside a loop are scheduled prior to the end of the loop.

> *Superstate I/O mode dictates that signal write operations be placed in the last c-step of a superstate. To avoid contradictory requirements, superstates that contain assignments to signals must stop at loop boundaries. If they did not, the rules of superstate I/O mode would attempt to move the write operations into (or even past) the loop. This means that a superstate cannot extend into a loop, nor can it extend beyond the end of the loop.*
>
> (Section 7.3.4)

This requirement translates into restrictions on the behavioral description. These rules are:

- Signal write operations cannot be moved *into* a loop.

    Signal writes that *precede* a loop must be separated from the *body* of the loop by at least one WAIT statement.

- Signal write operations cannot be moved *over* a loop.

    Signal writes that *precede* a loop must be separated from statements *after* the loop by at least one WAIT statement.

- Signal write operations cannot be moved *out* of a loop.

    Signal writes *inside* a loop must be separated from statements *after* the loop by at least one WAIT statement. EXIT statements must be considered when satisfying this rule.

- Signal write operations cannot be moved from the *bottom* of a loop to the *top* of the loop.

    Signal writes that occur after the *last* WAIT statement (i.e. in the last superstate) in the body of a loop must be separated from operations at the start (i.e. before the first WAIT statement) of the loop by at least one WAIT statement.

These rules are very similar to the restrictions associated with loops for fixed I/O mode. But note that these rules are based on the feature of superstate I/O mode that places all signal writes in the last c-step of the superstate. If there are only writes to variables, it is not necessary to include the WAIT statements.

Figure 13-17 shows different loop examples to be scheduled in superstate I/O mode. In the first two examples, there are no signal assignments inside the loop nor in the superstate that precedes the loop. Since there are no signal assignments that must be isolated from the loop, it is not necessary to add WAIT statements. In the last loop, the assignment must be isolated from the loop.

```
ready_for_input <= '0';
WAIT UNTIL clk'EVENT AND clk = '1';
-- Initialize internal variables
tmp1 := ( OTHERS => '0' );
tmp2 := x * y;

-- Perform matrix multiplication.
mult1:
    FOR i IN 0 TO 7 LOOP
        FOR j IN 0 TO 7 LOOP
            tmp := ( OTHERS => '0' );
            FOR k IN 0 TO 7 LOOP
                tmp := tmp + a( i * 8 + k ) * b( j * 8 + k );
            END LOOP;
            temp( i * 8 + j ) := tmp;
        END LOOP;
    END LOOP;
```

```
shift:
FOR i IN window_length - 1 DOWNTO 1 LOOP
    survivor_window( i ) := survivor_window( i - 1 );
END LOOP;

data_out <= branch_direction;
```
◄ Signal assignment does not interfere with the loop

```
ready_for_input <= '1';

wait_load: WHILE ( load /= '1' ) LOOP
    WAIT UNTIL clk'EVENT AND ( clk = '1' );
    EXIT reset_loop WHEN ( rst = '1' );
END LOOP;

ready_for_input <= '0';
```
◄ Signal assignment must be isolated from the loop

Figure 13-17: Loops to be scheduled in superstate I/O mode

The loop can be isolated from signal assignments with the following coding style:

- Start the loop with a WAIT statement.
- Place a WAIT statement immediately following the end of the loop.

This can also be accomplished with an alternate coding style:

- Place a WAIT statement immediately prior to the start of the loop.
- Place a WAIT statement immediately before all NEXT and EXIT statements.
- End the loop with a WAIT statement.

These coding styles are the same as those required for loops scheduled in fixed I/O mode (illustrated in Figures 13-13 and 13-14). The first coding style (shown in Figure 13-13) may be preferred, as it requires fewer WAIT statements that could be accidentally omitted.

In superstate I/O mode, WAIT statements are only required to isolate *signal assignments* from the loop. Variable assignments do not need to be isolated.

13.7 Free Scheduling Mode

Free mode is the least restrictive scheduling mode used by behavioral synthesis tools. In free mode, synthesis is totally free to add or delete clock cycles and to move I/O operations. There is one exception to this: when a signal is written to more than once, the order of those write operations are preserved.

There are very few coding style recommendations when using free I/O mode. However, in order to construct a synchronous behavioral description and to provide a structure for specifying reset behavior, the following coding style is recommended:

- Place reset actions at either the start or end of the process. The location of the reset actions was discussed in Section 13.2.2.
- Begin the main loop with a WAIT statement. This reflects the actual behavior of the design coming out of reset. After a reset, no assignments are made until the next clock edge.
- Create variables that correspond to the signals that will be read from and written to in the process. Whenever a signal is read from or written to, an I/O operation is generated. Each of these I/O operations are scheduled – behavioral synthesis assumes the values could be different. Signals should only be read from or written to multiple times if, in fact, the values can be different.

The application of the coding style to the simple algorithm is shown in Figure 13-18.

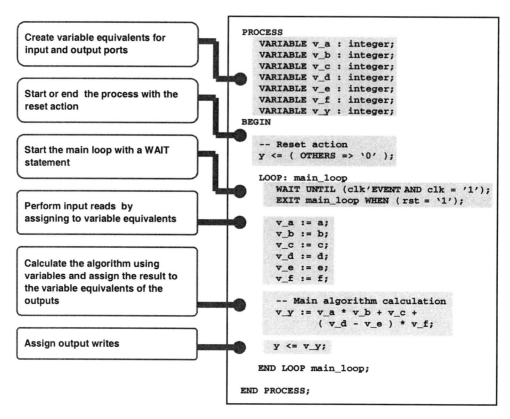

Figure 13-18: Simple algorithm coded for free I/O mode

For this simple design, the addition of the variables is unnecessary, but the code shows the overall structure that would be used for a more complex design.

13.8 Summary

Factors such as ease-of-development, understandability, and ability to perform architectural exploration can influence the coding style used for any particular design.

The I/O scheduling mode greatly affects the restrictions that are place on behavioral descriptions. There are rules for each I/O mode that influence how certain constructs (such as conditional statements and loops) can be described.

The way in which behavioral descriptions are written influences the quality of results. However, the way in which the designer directs the behavioral synthesis tool is equally important. Unlike in RTL design, a VHDL description does not imply a single architecture. A single behavioral description can be used to produce a number of design with a wide range of area and timing characteristics. For this reason, behavioral descriptions have fewer coding restrictions than RTL descriptions.

Using a particular behavioral description, a designer directs the behavioral synthesis tool to produce results based on the design goals. The number and type of components, the way in which arrays are mapped to memories, and the way loops are unrolled are but a few factors that can dramatically influence (positively *or* negatively) the architectures that are produced from a behavioral description.

Chapter 14

Case Study: JPEG Compression

14.1 Introduction

This chapter discusses an example design that implements a JPEG compression algorithm. JPEG is a compression standard developed for grayscale and color continuous-tone still images.

At the heart of the JPEG algorithm is a discrete cosine transform (DCT). This makes this design particularly exemplary, as this transformation is common in video and audio processing algorithms.

The chapter first discusses the algorithm. It then discusses how the algorithm can be coded in a behavioral style and the synthesis issues that should be considered in order to produce a quality result.

14.2 The Algorithm

There are two general categories of compression techniques: *lossless* and *lossy*.

As the name implies, lossless compression means that after compression and decompression, the output data is *exactly* the same as the input data: no data is lost during this process. Lossy compression, on the other hand, means that a certain amount of data is "thrown away" during compression or decompression: the output data is not necessarily the same as the input data.

Lossy compression is useful for video and audio application in which (typically) a certain amount of data loss is tolerable. The advantage of such techniques is that they can achieve very high compression rates, even if at the loss of some data.

JPEG compression is a lossy data compression algorithm. The algorithm allows a tradeoff between the amount of compression and the amount of data lost during compression and decompression.

The JPEG standard specifies a "baseline" method for encoding output data. The standard allows for modification of this method, depending upon the application. This example uses an encoding method that is best suited for a high-speed hardware implementation.

14.2.1 Algorithm Overview

Figure 14-1 shows a high level diagram of the algorithm.

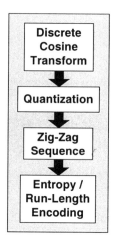

Figure 14-1: The JPEG Compression Algorithm

The first step in the algorithm is a discrete cosine transform (DCT). The DCT is similar to a Fast Fourier Transform (FFT): the DCT transforms points from the spatial domain to the "frequency" domain.

The DCT transforms a matrix of input pixels into an "average" value and smaller and smaller "frequency" components in a manner similar to the FFT. The DCT in the JPEG algorithm is processing two-dimensional input data.

The result of the DCT is another matrix. In this matrix, points further away from the origin [0,0] contribute less to the image.

The second step in the algorithm is quantization. Since points further away from the origin contribute less to the makeup of the image, certain points can be "thrown away" with little impact on the quality of the image. The process by which some points are discarded is called quantization. Since data is intentionally discarded, this step is a lossy transformation. (Other than round-off error that can occur from a digital implementation of the DCT, the DCT is lossless.)

The third step in the algorithm transforms a two-dimensional matrix into a linear sequence. The elements in the array are ordered using a "zigzag" sequence. This is done to maximize the number of consecutive zeros, which in turn allows for greater compression.

The final step in the algorithm performs the compression. This compression technique performs the following:

- Represents consecutive zeros in a small number of bits (since strings of zeros will be very common).

Case Study: JPEG Compression

- Uses fewer bits to represent more "likely" values.
- Indicates "end of block" if all remaining matrix values are zero.

Each of these steps is discussed in greater detail in the sections that follow.

14.2.2 Processing an Entire Image

This algorithm does not process an entire design at once. Rather, the design is divided into regions that are 8 pixels by 8 pixels, as shown in Figure 14-2.

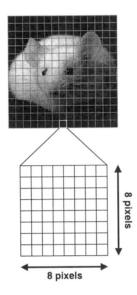

Figure 14-2: Images are divided into 8 x 8 matrices for compression

In this implementation of the algorithm, each pixel has a value from 0 to 255. The test bench that accompanies this design processes grayscale images, however, this algorithm could be used for color images as well. To process such images, the image would be separated into components (such as RGB or YCrCb), and each would be compressed separately.

14.2.3 The Discrete Cosine Transform

The formula for the discrete cosine transform is shown in Figure 14-3.

In this implementation, the input is an 8 x 8 matrix of grayscale values of the range 0 to 255. This range of values can be represented in 8 bits.

The transform requires 2 matrix multiplication operations. The input matrix is multiplied by a matrix of coefficients and that result is then multiplied by the transform of the matrix of coefficients. (The transform of a matrix is produced by taking the original matrix and swapping row positions with column positions).

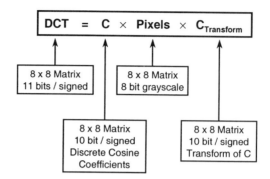

Figure 14-3: The discrete cosine transform (DCT)

The coefficient matrix is computed from the equation shown in Figure 14-4. The value of each element in the matrix is greater than −0.5 but less than 0.5. Each value is represented as a 10-bit signed value, with the binary point to the left of the first bit.

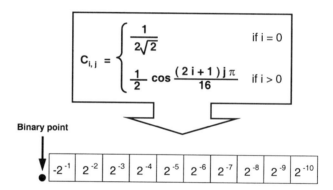

Figure 14-4: DCT Coefficients

For example, the value −0.125 is represented as:

$$-0.125 = -0.5 + 0.25 + 0.125$$
$$= -2^{-1} + 2^{-2} + 2^{-3}$$
$$= 1110000000$$

If greater (or lesser) precision were required, the size of this vector could be adjusted appropriately. Even though some data is lost with a 10-bit representation for the cosine coefficients, the results still seem to be of reasonable quality.

Except for the loss of data due to round-off error, the resulting matrix contains all of the information of the input matrix. This is a lossless transformation.

Case Study: JPEG Compression   249

Figure 14-5 provides a quick review of a matrix multiplication operation.

$$\begin{bmatrix} y_{0,0} & y_{0,1} \\ y_{1,0} & y_{1,1} \end{bmatrix} = \begin{bmatrix} a_{0,0} & a_{0,1} \\ a_{1,0} & a_{1,1} \end{bmatrix} \times \begin{bmatrix} b_{0,0} & b_{0,1} \\ b_{1,0} & b_{1,1} \end{bmatrix}$$

$$y_{0,0} = a_{0,0} * b_{0,0} + a_{0,1} * b_{1,0}$$
$$y_{0,1} = a_{0,0} * b_{0,1} + a_{0,1} * b_{1,1}$$
$$y_{1,0} = a_{1,0} * b_{0,0} + a_{1,1} * b_{1,0}$$
$$y_{1,1} = a_{1,0} * b_{0,1} + a_{1,1} * b_{1,1}$$

Figure 14-5: Matrix Multiplication

14.2.4 Quantization

Since the result of the DCT is stored in an 8 x 8 matrix of 11-bit signed values, the size of the data has actually increased from the original input!

However, much of this information will be discarded during quantization. Since points further away from the origin contribute less to the makeup of the image, certain points can be "thrown away" with little impact on the quality of the image. This step is a lossy transformation.

Quantization is performed by dividing each value in the matrix by a coefficient which varies depending upon its distance from the origin ([0,0]) and a quality factor which determines how much data to discard.

The quantization coefficient matrix is calculated using the equation in Figure 14-6. This

$$\text{Quantized Value}_{i,j} = \frac{DCT_{i,j}}{1 + ((1 + i + j) * \text{quality})}$$

For quality factor 2:

$$\begin{bmatrix} \frac{1}{3} & \frac{1}{5} & \frac{1}{7} & \frac{1}{9} & \frac{1}{11} & \frac{1}{13} & \frac{1}{15} & \frac{1}{17} \\ \frac{1}{5} & \frac{1}{7} & \frac{1}{9} & \frac{1}{11} & \frac{1}{13} & \frac{1}{15} & \frac{1}{17} & \frac{1}{19} \\ \frac{1}{7} & \frac{1}{9} & \frac{1}{11} & \frac{1}{13} & \frac{1}{15} & \frac{1}{17} & \frac{1}{19} & \frac{1}{21} \\ \frac{1}{9} & \frac{1}{11} & \frac{1}{13} & \frac{1}{15} & \frac{1}{17} & \frac{1}{19} & \frac{1}{21} & \frac{1}{23} \\ \frac{1}{11} & \frac{1}{13} & \frac{1}{15} & \frac{1}{17} & \frac{1}{19} & \frac{1}{21} & \frac{1}{23} & \frac{1}{25} \\ \frac{1}{13} & \frac{1}{15} & \frac{1}{17} & \frac{1}{19} & \frac{1}{21} & \frac{1}{23} & \frac{1}{25} & \frac{1}{27} \\ \frac{1}{15} & \frac{1}{17} & \frac{1}{19} & \frac{1}{21} & \frac{1}{23} & \frac{1}{25} & \frac{1}{27} & \frac{1}{29} \\ \frac{1}{17} & \frac{1}{19} & \frac{1}{21} & \frac{1}{23} & \frac{1}{25} & \frac{1}{27} & \frac{1}{29} & \frac{1}{31} \end{bmatrix}$$

Figure 14-6: Quantization Matrix

figure also shows coefficient values for a quality factor of 2. Note that a higher quality factor leads to discarding more data (and thus a more degraded final decompressed image). Even a quality factor of 2 can lead to compression ratios that approach 90%.

Each value is truncated after division. This means that many values in the matrix will have a value of zero after quantization. This provides a good opportunity for compression.

14.2.5 Zigzag Sequence

After quantization, many values in the matrix have a value of zero. The values further away from the origin ([0,0]) are more likely to be zero.

Much of the compression in this algorithm comes from a technique called *run-length encoding* (sometimes referred to as *recurrence coding*). This technique takes a number of recurring characters and represents them as the character, followed by a count.

Since quantization will lead to a matrix with many zeros, run-length encoding can greatly reduce the size of the image data. This technique will produce better results if more zeros occur in sequence.

Since the locations further away from the origin are more likely to be zero, the number of consecutive zeros is likely to be greater if the values are examined in a "zigzag" sequence as shown in Figure 14-7. This sequence is better than just examining the matrix by row or column.

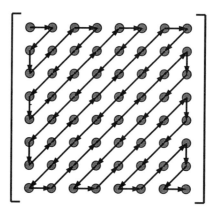

Figure 14-7: Zigzag sequence to maximize the number of consecutive zeros

The zigzag sequence begins in the upper-left corner of the matrix, and then proceeds in a zigzag pattern through the matrix. Note that values that are equally distant from the origin will appear next to one another in the output sequence.

This sequence of data is then compressed to produce the final compressed representation of the image.

14.2.6 Run-Length / Entropy Encoding

This algorithm uses three different encoding techniques: run-length encoding, entropy encoding, and "end of block" tagging.

As previously mentioned, run-length encoding converts a number of recurring characters into a character, followed by a count.

Entropy encoding is a compression technique in which more common patterns are represented by fewer bits than less common patterns.

After quantization, many values in the matrix have a value of zero. In addition, the values which are not close to the origin which are not zero will still tend to be close to zero. Thus, small values are more common than large values. To take advantage of this, small values are represented in fewer bits than large values, as shown in Figure 14-8.

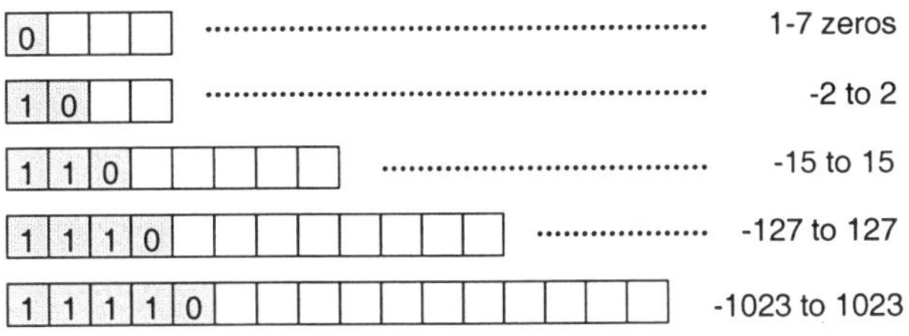

Figure 14-8: Output encoding

To simplify coding this algorithm in VHDL, matrix values are represented in one of five possible formats.

In the first format, the first bit is a zero. When the first bit is a zero, the three bits that follow represent a count value that ranges from 0 to 7. The bit pattern "0000" has a special meaning, indicating the "end of block" (i.e. all remaining matrix values are zero). Thus, the bit pattern "0100" represents a sequence of 4 zeros.

The remaining formats are used to represent non-zero values. Since smaller values are more likely to occur in the output stream, these values are represented with fewer bits. For example, the value −1 can be represented with 4 bits, but the value 299 requires 16 bits. Although it might seem that this approach should actually *increase* the size of the data, it does not because large values are very uncommon.

In the second format, the first two bits are "10". In this case, the following two bits represent the values −2, −1, 1, and 2. Note that the value 0 is already represented by the previous format. For this to work, the values are mapped to bit patterns as shown in Figure 14-9.

The remanding formats are used to represent the other ranges. All of the formats were deliberately specified to fall on *nibble* (4-bit) boundaries to simplify the coding of this algorithm in VHDL.

Value	Bit Representation
-2	00
-1	11
1	01
2	10

Figure 14-9: Mapping values in the range –2 to 2

The algorithm employs one additional method to increase compression. Since the largest values are found at [0,0] and these values tend to be similar from block to block, the value at [0,0] is not written out. For the value at [0,0] only, the difference between the value and value from the previous 8 x 8 block is output.

Sample output data is shown in Figure 14-10.

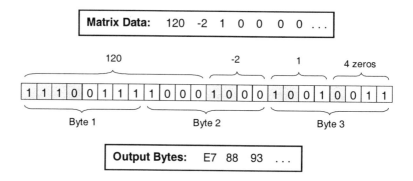

Figure 14-10: Relationship between matrix data and output bytes

14.3 The Environment

The design was developed with a test bench and test data. The test bench reads a "raw" grayscale image from a file and passes 8 x 8 portions of the image to the design. The size of the test image is specified using generics in the test bench. The default image size is 128 x 128 which means that there are 16 x 16 (or 256) matrices for each test image. The test bench takes the results of compressing each matrix and writes that data to a secondary file.

The test bench also incorporates a decompress design which reverses the compression process to reconstruct the image from the compressed data.

The final decompressed image is also written out to a file for later visualization.

Case Study: JPEG Compression

The overall structure of the test bench is shown in Figure 14-11.

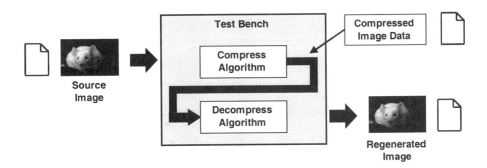

Figure 14-11: The design environment

The test environment includes a TCL script to display the image and statistics as the simulation is running.

Finally, there are two "stand-alone" TCL scripts which can be used to display "raw" images at actual size and 2x size, respectively.

A screen shot of a simulation run is shown in Figure 14-12.

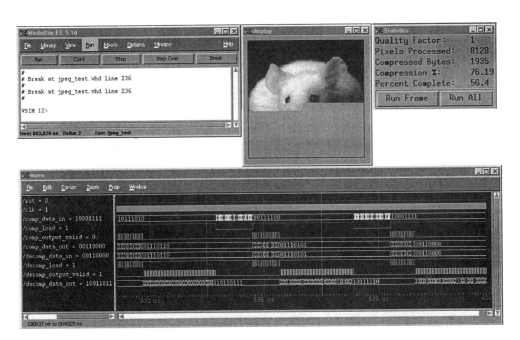

Figure 14-12: Sample simulation

14.4 Compression Results

Even though the algorithm does not implement the "baseline" encoding method of the JPEG compression standard, it still produces reasonable compression results.

The amount of compression that results from using different "quality" values is shown in Figure 14-13. Recall that the test images are 128 x 128 pixels, which means that the input file is 16,384 bytes.

"quality" Value	Compressed Size	Percentage of Original
1	3234	19.7 %
2	2054	12.5 %
4	1350	8.2 %
8	916	5.6 %

Figure 14-13: Compression statistics

The images begin to "break up" when the quality value is 4 or greater. Given that the perceived quality of an image is a subjective measure, it is impossible to say what quality value is "best" – it depends upon the application. The quality factor allows a tradeoff between the amount of compression and the fidelity of the reconstructed image. Based on the results of processing this particular image and other similar images, it seems that a quality factor of 2 produces reasonable compression without significant degradation in the quality of the image.

14.5 Behavioral Description

The creation of the behavioral description from the algorithm is quite straightforward. In this example, a package was created to store type, constant, and function declarations.

The package contains:

- the type definitions for the arrays that will hold the pixel, intermediate, and transformed matrices.
- a constant array which defines the DCT coefficients.
- the definition of a divide function, which is used during quantization.

The design itself is defined in a separate source file.

The entire source for this example can be found in Appendix A as well as on the CD that accompanies this book. Only essential portions of the source code are duplicated in this chapter.

14.5.1 Memories

The design defines three arrays that are used during the compression process. Theses arrays store the input pixel data (**pixels**), the temporary results after the first matrix multiplication (**temp**), and the results of the second matrix multiplication (**dct**). Each of these arrays represent 8 x 8 matrices. These arrays are all mapped to memories to avoid the significant cost of multiplexing unmapped arrays (as discussed in Chapter 4).

Case Study: JPEG Compression

14.5.2 Handshaking

The implementation of the JPEG compression algorithm uses full handshaking to read input data and write output data. The algorithm begins by reading 64 input values that represent an 8 x 8 image. The image is then compressed. Once complete, the design uses an output handshaking protocol to write out the compressed bytes.

14.5.3 Matrix Multiplication

The heart of the JPEG compression algorithm is the discrete cosine transform. This transform is composed of two, 8 x 8 matrix multiplications. This portion of the design contributes a large portion of the area cost because it involves large multiplications. Also, the memories are significantly exercised during these computations. Finally, the structure of a matrix multiplication lends itself to varying implementations.

The code that implements the first matrix multiplication is shown in Figure 14-14.

```
mult1: FOR i IN 0 TO 7 LOOP
         FOR j IN 0 TO 7 LOOP
            tmp := ( OTHERS => '0' );
            FOR k IN 0 TO 7 LOOP
               tmp := tmp + pixels( i * 8 + k ) * c( j * 8 + k );
            END LOOP;
            temp( i * 8 + j ) := tmp;
         END LOOP;
      END LOOP;
```

Figure 14-14: Code to perform first matrix multiplication

In this case study, the design is scheduled using superstate I/O mode. Because the loops contain only variable assignments, it is not necessary to place WAIT statements in or around the loop.

Note that the innermost loop assigns to a temporary variable and then *after the loop*, assigns to the array **temp**. It might seem more natural to assign to the temporary array inside this loop. However, every assignment to the array **temp** corresponds to a memory write, which must be scheduled. The number of memory write operations can be reduced by using a temporary variable.

14.6 Behavioral Synthesis

The results of behavioral synthesis can be greatly affected by the resource and time constraints specified by the user. The manner in which variables are mapped to memories and how loops are unrolled (or left rolled) can have at least as much impact.

Consider again the first matrix multiplication shown in Figure 14-14. It may seem reasonable to unroll all the loops and let the scheduler determine the amount of parallelism that is possible. However, note that the inner loop contains a memory read operation, a multiply operation, and an add operation. If the loops are unrolled completely, the are 8 x 8 x 8 = 256 copies of these operations that must be scheduled. While this may not prevent the design from being scheduled, it will certainly slow down processing and the sheer number of operations will make the results of synthesis difficult to understand.

Consider how the implementation will operate. First, a value is read from the **pixel** memory. This value is multiplied by a constant value and added to the running sum. This read / multiply / accumulate is performed 8 times. The accumulated result is written to the **temp** memory.

If all three loops are left rolled, there is no opportunity for parallelism between multiple iterations of the loop. But the memory read and multiply operations are independent. The only the data dependency from one iteration of the loop to the next is associated with the accumulate operation.

To explore this possible parallelism, the two outer loops are left rolled, but the inner loop is unrolled. The pragmas that specify the unrolling are shown in the code in Figure 14-15.

```
mult1: FOR i IN 0 TO 7 LOOP    -- pragma dont_unroll
           FOR j IN 0 TO 7 LOOP    -- pragma dont_unroll
               tmp := ( OTHERS => '0' );
               FOR k IN 0 TO 7 LOOP
                   tmp := tmp + pixels( i * 8 + k ) * c( j * 8 + k );
               END LOOP;
               temp( i * 8 + j ) := tmp;
           END LOOP;
       END LOOP;
```

Figure 14-15: Code to perform first matrix multiplication with pragmas to control unrolling

14.6.1 Bounding the Design Space

When exploring architectures using a behavioral synthesis tool, it is often useful to understand the extreme points in the design space. One of these points is described as the "fastest" implementation, which is the most parallel implementation using any number of the fastest components in the target technology library. Another point of interest is described as the "smallest" implementation, which is the implementation which uses the minimum number of the smallest components in the target technology library. The name "smallest" is somewhat misleading, because it refers to the size of the data path portion of the design. When fewer resources are used, the number of multiplexers and registers required for sharing increases. So, the final area numbers should be compared.

The architectures that are discussed in this chapter were generated using Monet®, the behavioral design environment from Mentor Graphics Corporation. The loop report for the "fastest" implementation is shown in Figure 14-16.

This report shows both the number of c-steps *and* the number of clock cycles required to execute the loops. The loop that represents the first matrix multiplication (**mult1**) contains the loop **loop122**. There is no loop inside **loop122**, since it was completely unrolled. The loop **loop122** was scheduled in 6 c-steps. Since this loop is executed 8 times, the loop requires 6 x 8 = 48 clock cycles to complete. The outer loop, **mult1**, requires one additional c-step (it is scheduled in 7 c-steps), and is also executed 8 times. This means that the loop **mult1** requires (48 + 1) * 8 = 392 clock cycles to complete. This number is reflected in the report.

Case Study: JPEG Compression

```
                                  Itera-    C-Steps
       Loop                       tions  Len (From:To)  Cycles
       ---------------------      -----  -------------  ------
       main_loop                    1    31  (1:31)      1872
         read_loop                 64     3  (1:3)        192
           wait_load                1     1  (1:1)          1
         mult1                      8     7  (4:10)       392
           loop122                  8     6  (4:9)         48
         mult2                      8    13  (11:23)      776
           loop139                  8    12  (11:22)       96
         generate_results          64     8  (24:31)      512
           nibble_loop              1     4  (27:30)        4
             out_rdy_loop           1     1  (27:27)        1
```

Figure 14-16: Loop report for the "fastest" implementation

The area report for the *data path* portion of this design is shown in Figure 14-17.

```
Qualified Component Name            Area      Quantity    Area Subtotal
----------------------------       --------   --------    -------------
AndGate(1,2)                          5.00       1             5.00
AndGate(1,3)                          6.00       2            12.00
AndGate(1,6)                         12.00       2            24.00
InvGate(1)                            3.00       1             3.00
NorGate(1,6)                         14.00       2            28.00
OrGate(1,2)                           5.00       1             5.00
XorGate(1,2)                         10.00       1            10.00
abs_cla(1,11)                       415.52       1           415.52
abs_cla(1,8)                        276.98       1           276.98
add_bki(0,4,4)                      134.00       2           268.00
add_bki(0,4,5)                      166.12       2           332.25
add_bki(0,4,6)                      193.74      13          2518.63
add_vn(1,19,21)                    1076.06       3          3228.18
add_vn(1,31,31)                    1637.00       3          4911.00
cmp_cla(0,1,1,1,6,6)                157.78       1           157.78
cmp_cla(0,1,1,1,8,9)                228.60       3           685.80
cmp_cla(1,1,1,1,4,11)               277.40       1           277.40
cmp_cla(1,1,1,1,9,11)               293.94       2           587.87
inc_vn(0,4)                          53.00      15           795.00
inc_vn(0,6)                          89.52       2           179.04
inc_vn(0,8)                         131.82       1           131.82
mult_bth_cla(1,10,21,31)           5609.96       7         39269.70
mult_bth_cla(1,9,10,19)            2484.71       7         17393.00  <--
mult_wallace_cla(0,4,4,8)           361.00       1           361.00
mux_mux2(64,10)                    3036.56       8         24292.50
mux_mux2(64,6)                     1882.94       1          1882.94
ram_s_RW_port(1,64,9)                 0.00       7             0.00  <--
ram_s_RW_port(2,64,21)                0.00       7             0.00
ram_s_RW_port(3,64,11)                0.00       1             0.00
rdcand_two(1,4)                      12.00       1            12.00
rdcnor_two(4,11)                     39.19       2            78.38
rdcor_four(3,6)                      11.82       1            11.82
rdcor_three(3,11)                    26.24       1            26.24
rdcor_three(3,12)                    28.00       1            28.00
rdcor_two(3,4)                       12.00       1            12.00
sub_bki(1,11,11)                    466.00       2           932.00
umin_cla(1,11)                      322.29       1           322.29
write_indexB(1,0)                     0.00       3             0.00
ram_s_nRW(1,64,9,7)                   0.00       1             0.00
ram_s_nRW(2,64,21,7)                  0.00       1             0.00
ram_s_nRW(3,64,11,1)                  0.00       1             0.00
                                                           -----------
                                         Total Area:       99474.10
```

Figure 14-17: Data path area report for the "fastest" implementation

Note the two components that are highlighted in the report, a multiplier and a RAM port that are used in the first matrix multiplication. There are 7 instances of each of these components in the design. The number of these components is large because the solution was generated without any restriction on the resources. This report indicates that if 7 multipliers were available and a RAM with 7 ports were available, they would all be used concurrently at some point in the calculation. These large numbers of components can be used concurrently because the multiply operations in the matrix multiplications can be calculated at the same time.

While this solution does indicate something about the parallelism that is possible when implementing this design, the large number of multipliers and the large number of ports on the RAM make this implementation impractical.

Consider now the "smallest" implementation, which is the implementation which uses the minimum number of the smallest components in the target technology library. The loop report for the "smallest" implementation is shown in Figure 14-18.

```
                       Itera-    C-Steps
Loop                   tions   Len (From:To)   Cycles
-----------------------------------------------------
main_loop                1      95  (1:95)      5968
  read_loop             64       3  (1:3)        192
    wait_load            1       1  (1:1)          1
  mult1                  8      30  (4:33)      1864
    loop122              8      29  (4:32)       232
  mult2                  8      48  (34:81)     3016
    loop139              8      47  (34:80)      376
  generate_results      64      14  (82:95)      896
    nibble_loop          1       4  (91:94)        4
      out_rdy_loop       1       1  (91:91)        1
```

Figure 14-18: Loop report for the "smallest" implementation

From the report, it is obvious that this implementation takes many more clock cycles to perform the calculation than the "fastest" implementation. The first matrix multiplication takes almost five times the number of *clock cycles* (1864 clock cycles vs. 392 clock cycles). But note that the difference in the number of *c-steps* required to schedule loop **loop122** is only 23 (30 c-steps vs. 7 c-steps).

> The distinction between c-steps and clock cycles is important when comparing schedules. While the overall length of the schedules in c-steps can give some insight into the differences between the schedules, the comparative number of clock cycles that are required to complete the calculation is what is most important.

Case Study: JPEG Compression

The area report for the data path portion of the "smallest" design is shown in Figure 14-19.

```
Qualified Component Name              Area      Quantity   Area Subtotal
--------------------------------      --------  --------   -------------
AndGate(1,2)                             5.00       1           5.00
AndGate(1,3)                             6.00       2          12.00
AndGate(1,6)                            12.00       1          12.00
InvGate(1)                               3.00       1           3.00
NorGate(1,6)                            14.00       2          28.00
OrGate(1,2)                              5.00       1           5.00
XorGate(1,2)                            10.00       1          10.00
abs_rpl(1,8)                           161.00       1         161.00
abs_rpl1(1,11)                         234.29       1         234.29
add_rpl(0,4,4)                          72.00       1          72.00
add_rpl(0,4,5)                          84.36       1          84.36
add_rpl(0,4,6)                          96.77       1          96.77
add_rpl(1,19,21)                       417.56       1         417.56
add_rpl(1,31,31)                       591.00       1         591.00
cmp_cla(0,1,1,1,4,4)                   102.00       1         102.00
cmp_rpl1(0,1,1,1,6,6)                  128.11       1         128.11
cmp_rpl1(0,1,1,1,8,9)                  167.77       1         167.77
cmp_rpl1(1,1,1,1,4,11)                 202.04       1         202.04
cmp_rpl1(1,1,1,1,9,11)                 235.84       1         235.84
inc_rpl(0,4)                            36.00       1          36.00
inc_rpl(0,6)                            58.00       1          58.00
inc_rpl(0,8)                            80.00       1          80.00
mult_bth_rpl(1,10,21,31)              5133.70       1        5133.70
mult_bth_rpl(1,9,10,19)               2158.28       1        2158.28   <---
mult_wallace_rpl(0,4,4,8)              275.77       1         275.77
mux_mux2(64,10)                       3036.56       3        9109.68
mux_mux2(64,6)                        1882.94       1        1882.94
ram_s_RW_port(1,64,9)                    0.00       1           0.00   <---
ram_s_RW_port(2,64,21)                   0.00       1           0.00
ram_s_RW_port(3,64,11)                   0.00       1           0.00
rdcand_two(1,4)                         12.00       1          12.00
rdcnor_two(4,11)                        39.19       2          78.38
rdcor_four(3,6)                         11.82       1          11.82
rdcor_three(3,11)                       26.24       1          26.24
rdcor_three(3,12)                       28.00       1          28.00
rdcor_two(3,4)                          12.00       1          12.00
sub_rpl(0,9,8)                         203.05       1         203.05
sub_rpl(1,11,11)                       249.00       1         249.00
sub_vni(0,4,4)                          81.00       1          81.00
umin_rpl(1,11)                         145.00       1         145.00
write_indexB(1,0)                        0.00       1           0.00
ram_s_nRW(1,64,9,1)                      0.00       1           0.00
ram_s_nRW(2,64,21,1)                     0.00       1           0.00
ram_s_nRW(3,64,11,1)                     0.00       1           0.00
                                                         -------------
                                              Total Area:     22148.60
```

Figure 14-19: Data path area report for the "fastest" implementation

As expected, this implementation uses only a single multiplier and a single RAM port. But the reduction in data path area comes at a significant cost of latency.

14.6.2 Exploring Other Architectures

The "fastest" and "smallest" architectures provide some insight into the design space that can be explored for this algorithm. With this information, the design space can be constrained to produce some intermediate results.

Consider again the "fastest" implementation, but with an additional constraint. Assume that the target technology most easily supports RAMs with a single port. This resource constraint can be applied to the behavioral synthesis tool. The resulting loop and data path area reports are shown in Figure 14-20.

```
                                 Itera-    C-Steps
Loop                             tions   Len (From:To)   Cycles
------------------------------   ------  -------------   ------
main_loop                          1       40 (1:40)      2448
  read_loop                       64        3 (1:3)        192
    wait_load                      1        1 (1:1)          1
  mult1                            8       11 (4:14)       648
    loop122                        8       10 (4:13)        80
  mult2                            8       18 (15:32)     1096
    loop139                        8       17 (15:31)      136
  generate_results                64        8 (33:40)      512
    nibble_loop                    1        4 (36:39)        4
      out_rdy_loop                 1        1 (36:36)        1
```

```
Qualified Component Name              Area        Quantity    Area Subtotal
------------------------              --------    --------    -------------
AndGate(1,2)                             5.00        1             5.00
AndGate(1,3)                             6.00        2            12.00
AndGate(1,6)                            12.00        2            24.00
InvGate(1)                               3.00        1             3.00
NorGate(1,6)                            14.00        2            28.00
OrGate(1,2)                              5.00        1             5.00
XorGate(1,2)                            10.00        1            10.00
abs_cla(1,11)                          415.52        1           415.52
abs_cla(1,8)                           276.98        1           276.98
add_bki(0,4,4)                         134.00        2           268.00
add_bki(0,4,5)                         166.12        2           332.25
add_bki(0,4,6)                         193.74       13          2518.63
add_vn(1,19,21)                       1076.06        3          3228.18
add_vn(1,31,31)                       1637.00        3          4911.00
cmp_cla(0,1,1,1,6,6)                   157.78        1           157.78
cmp_cla(0,1,1,1,8,9)                   228.60        2           457.20
cmp_cla(1,1,1,1,4,11)                  277.40        1           277.40
cmp_cla(1,1,1,1,9,11)                  293.94        2           587.87
inc_vn(0,4)                             53.00       15           795.00
inc_vn(0,6)                             89.52        2           179.04
inc_vn(0,8)                            131.82        1           131.82
mult_bth_cla(1,10,21,31)              5609.96        1          5609.96
mult_bth_cla(1,9,10,19)               2484.71        1          2484.71
mult_wallace_cla(0,4,4,8)              361.00        1           361.00
mux_mux2(64,10)                       3036.56        8         24292.50
mux_mux2(64,6)                        1882.94        1          1882.94
ram_s_RW_port(1,64,9)                    0.00        1             0.00
ram_s_RW_port(2,64,21)                   0.00        1             0.00
ram_s_RW_port(3,64,11)                   0.00        1             0.00
rdcand_two(1,4)                         12.00        1            12.00
rdcnor_two(4,11)                        39.19        2            78.38
rdcor_four(3,6)                         11.82        1            11.82
rdcor_three(3,11)                       26.24        1            26.24
rdcor_three(3,12)                       28.00        1            28.00
rdcor_two(3,4)                          12.00        1            12.00
sub_bki(1,11,11)                       466.00        2           932.00
umin_cla(1,11)                         322.29        1           322.29
write_indexB(1,0)                        0.00        2             0.00
ram_s_nRW(1,64,9,1)                      0.00        1             0.00
ram_s_nRW(2,64,21,1)                     0.00        1             0.00
ram_s_nRW(3,64,11,1)                     0.00        1             0.00
                                                         -------------
                                              Total Area:    50677.50
```

Figure 14-20: Loop and data path area report for an architecture with 1 RAM port

Case Study: JPEG Compression

Consider a final scenario in which a 2-port RAM *could* be created for the target technology, but perhaps at additional cost. Behavioral synthesis tools allow the designer to understand the impact of such a decision on the area and delay of the implementation. The loop and data path area reports for the architecture that results from allowing an additional RAM port is shown in Figure 14-21.

```
                         Itera-    C-Steps
Loop                     tions  Len (From:To)  Cycles
-----------------------  ------ -------------  ------
main_loop                  1      35 (1:35)     2128
  read_loop               64       3 (1:3)       192
    wait_load              1       1 (1:1)         1
  mult1                    8       9 (4:12)      520
    loop122                8       8 (4:11)       64
  mult2                    8      15 (13:27)    904
    loop139                8      14 (13:26)    112
  generate_results        64       8 (28:35)    512
    nibble_loop            1       4 (31:34)      4
      out_rdy_loop         1       1 (31:31)      1
```

```
Qualified Component Name            Area      Quantity   Area Subtotal
--------------------------------   --------   --------   -------------
AndGate(1,2)                          5.00        1           5.00
AndGate(1,3)                          6.00        2          12.00
AndGate(1,6)                         12.00        2          24.00
InvGate(1)                            3.00        1           3.00
NorGate(1,6)                         14.00        2          28.00
OrGate(1,2)                           5.00        1           5.00
XorGate(1,2)                         10.00        1          10.00
abs_cla(1,11)                       415.52        1         415.52
abs_cla(1,8)                        276.98        1         276.98
add_bki(0,4,4)                      134.00        2         268.00
add_bki(0,4,5)                      166.12        2         332.25
add_bki(0,4,6)                      193.74       13        2518.63
add_vn(1,19,21)                    1076.06        3        3228.18
add_vn(1,31,31)                    1637.00        3        4911.00
cmp_cla(0,1,1,1,6,6)                157.78        1         157.78
cmp_cla(0,1,1,1,8,9)                228.60        3         685.80
cmp_cla(1,1,1,1,4,11)               277.40        1         277.40
cmp_cla(1,1,1,1,9,11)               293.94        2         587.87
inc_vn(0,4)                          53.00       15         795.00
inc_vn(0,6)                          89.52        2         179.04
inc_vn(0,8)                         131.82        1         131.82
mult_bth_cla(1,10,21,31)           5609.96        2       11219.90
mult_bth_cla(1,9,10,19)            2484.71        2        4969.42
mult_wallace_cla(0,4,4,8)           361.00        1         361.00
mux_mux2(64,10)                    3036.56        8       24292.50
mux_mux2(64,6)                     1882.94        1        1882.94
ram_s_RW_port(1,64,9)                 0.00        2           0.00
ram_s_RW_port(2,64,21)                0.00        2           0.00
ram_s_RW_port(3,64,11)                0.00        1           0.00
rdcand_two(1,4)                      12.00        1          12.00
rdcnor_two(4,11)                     39.19        2          78.38
rdcor_four(3,6)                      11.82        1          11.82
rdcor_three(3,11)                    26.24        1          26.24
rdcor_three(3,12)                    28.00        1          28.00
rdcor_two(3,4)                       12.00        1          12.00
sub_bki(1,11,11)                    466.00        2         932.00
umin_cla(1,11)                      322.29        1         322.29
write_indexB(1,0)                     0.00        3           0.00
ram_s_nRW(1,64,9,2)                   0.00        1           0.00
ram_s_nRW(2,64,21,2)                  0.00        1           0.00
ram_s_nRW(3,64,11,1)                  0.00        1           0.00
                                                          -------------
                                            Total Area:    59000.70
```

Figure 14-21: Loop and data path area report for an architecture with 2 RAM ports

The design characteristics of the four architectures that have been discussed are shown in Figure 14-22.

	"Fastest"	"Smallest"	1 RAM Port	2 RAM Ports
Data path area	~ 99 k	~ 22 k	~ 51 k	~ 59 k
Number of c-steps	31	95	40	35
Number of clock cycles	1872	5968	2448	2128

Figure 14-22: Comparison of the four architectures

With Monet®, these architectures were generated in a matter of minutes. In a small amount of time, a designer can explore a variety of possible implementations for a particular algorithm. This type of exploration is just simply not possible in a timely fashion using a traditional RTL design flow.

14.7 Summary

This case study illustrates that a relatively complicated algorithm can be implemented in a small amount of behavioral code. Excluding comments, the code required to describe this algorithm (including the code in the package), is less than 400 lines.

This is a particularly good example for showing the wide variation in architectures that can be produced by modifying the available resources. This is particularly applicable to this algorithm, that might be used in a digital camera.

It is easy to imagine that there are many target markets for such a camera. A high-end professional user would need to be able to take many pictures each second, which may require a very fast implementation of this algorithm. For such a market, the cost of the end product (which allows for higher silicon cost) is likely to be high. However, for the general consumer, a more modestly priced product (which means lower silicon costs) is probably required. The performance of the algorithm in this product may be less important.

In this case study, the design space was first bounded by analyzing the "fastest" and "smallest" designs. Then, a more realistic design that used a single RAM port was generated. Finally, an architecture that used 2 RAM ports was explored, to see the impact of the addition RAM port on the implementation.

This example highlights the reusability of behavioral code (the same code can be used to create implementations appropriate for very different applications). It also shows how behavioral synthesis technology supports "what-if" analysis (as shown through the analysis of an implementation that uses a 2-port RAM).

Chapter 15

Case Study: FIR Filter

15.1 Introduction

This chapter discusses an example design that implements a Finite Impulse Response (FIR) digital filter. Digital filtering is an important aspect of most DSP-oriented designs. This filter can be easily described using a behavioral coding style and the results of behavioral synthesis are easily understood.

The chapter first discusses the filter function. It then discusses how the filter can be coded in a behavioral style and the synthesis issues that should be considered in order to produce a quality result.

15.2 The Algorithm

There are two general categories of digital filters: *Finite Impulse Response (FIR)* and *Infinite Impulse Response (IIR)*. FIR filters can be implemented in a non-recursive manner and thus are stable. IIR filters are not always stable.

A FIR filter is described by the following equation:

$$y(n) = \sum_{i=0}^{t} h(i) * x(n - i)$$

This equation defines the output of the filter as a function of previous input values. The variable t represents the number of previous input values that are used in the calculation of the output value. This value represents the number of *taps* associated with the filter. The filter described in this example has 15 taps.

The variable x represents input data. The index associated with x indicates which sample of the input is being referred to. For example, $x(0)$ is the current input data; $x(1)$ is the previous input data, and so on.

The variable h represents the tap coefficients that define the particular characteristics of the filter. These values are constant for a particular filter implementation.

15.3 Behavioral Description

15.3.1 Tap Coefficients

The FIR filter described in this chapter is not "hard-coded" to a particular set of coefficients. Rather, the design can be used for any set of coefficients that define a 15-tap FIR filter.

This means that the design must include a mechanism for loading coefficient values. To accomplish this, the behavioral description has a "configuration loop" which is dedicated to loading tap coefficients. This loop is shown in Figure 15-1.

```
config_loop: LOOP

    --
    -- Wait until next clock cycle to check for
    -- load or start signals.
    --
    WAIT UNTIL  clk'EVENT  AND  ( clk = '1' );
    EXIT reset_loop WHEN ( rst = '1' );

    IF ( start = '1' ) THEN
        EXIT config_loop;
    END IF;

    IF ( load = '1' ) THEN
        coeff ( conv_integer( '0' & coeff_addr ) ) := data_in;
    END IF;

END LOOP config_loop;
```

Figure 15-1: Coefficient configuration loop

The loading of coefficients is performed using a load signal and a coefficient address. To load a particular coefficient value, the input signal **coeff_addr** is set to the coefficient address (0 to 14), the input signal **data_in** is set to the coefficient value, and the input signal **load** is set to '1' to indicate a valid coefficient value. This process can be repeated until all of the coefficient values have been loaded.

The signal **start** is used to exit the configuration loop. When this signal is set to '1', the configuration loop is exited, and filtering begins.

In this example, the coefficients are represented by 12 bits. The number of bits that are used to represent the coefficients has a direct effect on the precision of the filter. The coefficient values are signed and are between the values −1.0 and +1.0 (excluding these two values).

Similar to the DCT coefficients that were used in the JPEG case study, the binary point for the coefficient values appears to the left of the first digit.

15.3.2 Filtering

The coefficient configuration loop is followed by the loop that actually performs the filtering function. This loop is shown in Figure 15-2.

```
filter_loop: LOOP

    WAIT UNTIL clk'EVENT AND ( clk = '1' );
    EXIT reset_loop WHEN ( rst = '1' );

    history( 0 ) := signed( data_in );

    sum := ( OTHERS => '0' );
    FOR i IN 0 TO 14 LOOP
        sum := sum + history( i ) * coeff( i );
    END LOOP;

    data_out <= sum( 23 DOWNTO 12 );

    FOR i IN 14 DOWNTO 1 LOOP
        history( i ) := history( i - 1 );
    END LOOP;

END LOOP filter_loop;
```

Figure 15-2: Filter loop

The loop calculates a running sum of the product of each tap coefficient and its corresponding previous input value. The previous input values are stored in an array called **history**. Because the filter has 15 taps, 15 input values must be saved. Once the output value is calculated, it is assigned to the output signal **data_out**. Only a slice of variable **sum** is assigned to the signal **data_out**. This is done to realign the binary point. Recall that the variable **history** has 12 digits to the right of the binary point and that the variable **coeff** has 12 digits to the right of the binary point. After these variables are multiplied together, the result has 24 (i.e. 12 + 12) digits to the right of the binary point. Only a slice of the variable **sum** is used, discarding the 12 least significant bits.

Finally, the history values are shifted, in preparation for the next input value.

In this example, the input data (and thus each element of the history array) is represented by 12 bits. Just like the coefficients, the number of bits that are used to represent the input data also has a direct effect on the precision of the filter.

The code in Figure 15-2 does not represent an efficient implementation for the calculation of the value of **sum**. The use of a loop results in a long chain of cascaded adders. This chain is implied when the loop is unrolled. The original loop and the equivalent code after loop unrolling is shown in Figure 15-3. The code after loop unrolling implies the general structure shown in Figure 15-4.

```
sum := ( OTHERS => '0' );
FOR i IN 0 TO 14 LOOP
   sum := sum + history( i ) * coeff( i );
END LOOP;
```

Loop Unrolling

```
sum := ( OTHERS => '0' );
sum := sum + history( 0 ) * coeff( 0 );
sum := sum + history( 1 ) * coeff( 1 );
sum := sum + history( 2 ) * coeff( 2 );
   . . .
sum := sum + history( 12 ) * coeff( 12 );
sum := sum + history( 13 ) * coeff( 13 );
sum := sum + history( 14 ) * coeff( 14 );
```

Figure 15-3: Inner loop, before and after loop unrolling

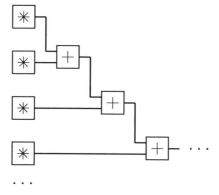

Figure 15-4: Implied structure of unrolled inner loop

The resultant chain of adders will likely result in a longer than necessary latency when the design is scheduled. Figure 15-5 shows an alternate description for the calculation of the output value. In this code, parenthesis are used to force the construction of a balanced tree of adders.

Case Study: FIR Filter

```
sum := ( ( ( history(  0 ) * coeff(  0 ) +
             history(  1 ) * coeff(  1 ) ) +
           ( history(  2 ) * coeff(  2 ) +
             history(  3 ) * coeff(  3 ) ) ) +
         ( ( history(  4 ) * coeff(  4 ) +
             history(  5 ) * coeff(  5 ) ) +
           ( history(  6 ) * coeff(  6 ) +
             history(  7 ) * coeff(  7 ) ) ) ) +
       ( ( ( history(  8 ) * coeff(  8 ) +
             history(  9 ) * coeff(  9 ) ) +
           ( history( 10 ) * coeff( 10 ) +
             history( 11 ) * coeff( 11 ) ) ) +
         ( ( history( 12 ) * coeff( 12 ) +
             history( 13 ) * coeff( 13 ) ) +
             history( 14 ) * coeff( 14 ) ) ) );

data_out <= sum( 24 DOWNTO 9 );
```

Figure 15-5: Alternate description for sum calculation

The entire source for this example can be found in Appendix B as well as on the CD that accompanies this book.

15.4 The Environment

The design was tested using a set of coefficients that implement a low-pass filter. The design was developed with a test bench. The test bench creates a sequence of sinusoidal waves that move from high to low frequency. Conversion functions from real numbers to bit vectors and from bit vectors to real numbers are used by the test bench to facilitate the viewing of the actual waveforms during simulation.

A screen shot of a simulation run is shown in Figure 15-6. The input and output waveforms show the expected response of a low-pass filter.

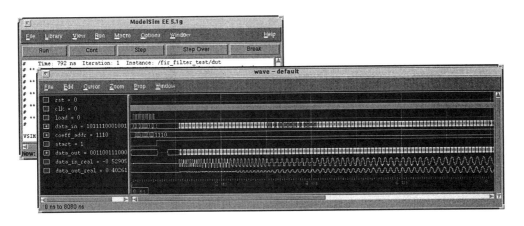

Figure 15-6: Simulation of the FIR filter

15.5 Behavioral Synthesis

15.5.1 Bounding the Design Space

As mentioned in the JPEG case study, it is often useful to understand the extreme points in the design space. So, for this design, the "fastest" and "smallest" implementations will be examined. To review, the "fastest" implementation is the most parallel implementation using any number of the fastest components in the target technology library and the "smallest" implementation is the implementation which uses the minimum number of the smallest components in the target technology library.

The architectures that are discussed in this chapter were generated using Monet®, the behavioral design environment from Mentor Graphics Corporation. The Gantt chart for the "fastest" implementation is shown in Figure 15-7.

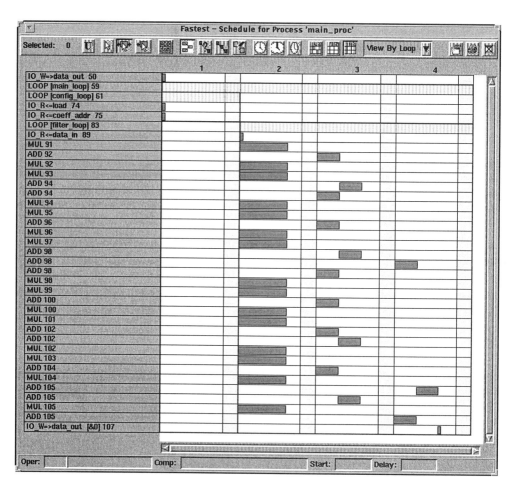

Figure 15-7: Gantt chart for "fastest" solution

Case Study: FIR Filter

This implementation requires 15 multipliers and 11 adders and is scheduled in 4 c-steps. The schedule shows that a fast multiplier can be scheduled in a single clock cycle and that two add operations can fit in a clock cycle.

But the filter could be implemented with a smaller number of components. In fact, the "smallest" implementation uses only a single multiplier component and a single adder component! Unfortunately, this comes at a high cost. The latency of the design is greatly extended. The "smallest" implementation uses a multiplier and an adder that are both multi-cycle components. This results in a schedule that is 46 c-steps long. The Gantt chart for the "smallest" solution is shown in Figure 15-8.

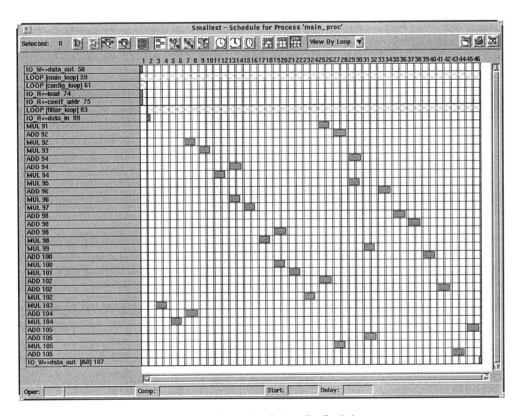

Figure 15-8: Gantt chart for "smallest" solution

Solutions with latencies between 4 and 46 c-steps can be achieved by varying the number and type of multipliers that are available.

While many different implementations are possible, the input data rate will determine what solutions are viable implementations of this filter.

Consider the "fastest" solution. The filter loop in this implementation is scheduled in 4 c-steps (the first c-step in the schedule is the configuration loop). This means that the input data is only sampled every 3 clock cycles. If the input data arrives at a faster rate, much of the data will be lost, resulting in incorrect output.

15.5.2 Pipelining the Design

Pipelining can be used to increase the throughput of the design. Assume that the data arrives at the same rate as the clock period. In order to process every piece of input data, the filter loop must either be scheduled in 1 c-step *or* the loop must be pipelined with an initialization interval of 1. The schedule length of the filter loop in the "fastest" solution is 3 c-steps, so pipelining must be considered.

Recall from the discussion in Chapter 8 how the filter loop can be pipelined. First, the loop is duplicated as many times as necessary to determine the maximum overlap. This is shown in Figure 15-9.

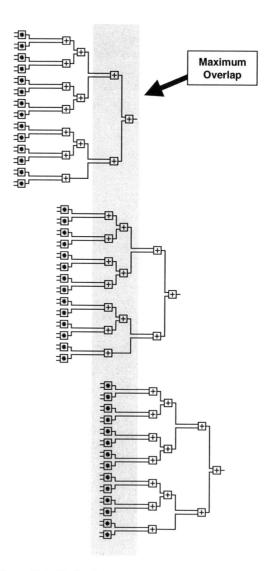

Figure 15-9: Pipelined design with an initialization interval of 1

Case Study: FIR Filter

This figure shows that the filter loop can be pipelined in 3 stages using 15 multiplier components and 14 adder components. The latency of the implementation is still 3 clock cycles, but the throughput has been tripled and only at a cost of 3 adder components!

Now consider a different input data rate, in which the data arrives every other clock cycle. The "fastest" implementation can only read input data every 3 clock cycles, so it cannot be used to meet this requirement.

Again, the design must be pipelined, but now with an initialization interval of 2. This relaxes the constraints on the resources. In the previous pipelined implementation, all of the multiplication operations had to be performed in a single clock cycle. With an initialization interval of 2, the multiplications can be spread out over the two clock cycles, reducing the number of multiplier components required. The design, pipelined with an initialization interval of 2, is shown in Figure 15-10. Note that this implementation is implemented in 2 stages and requires only 8 multiplier components and 7 adder components.

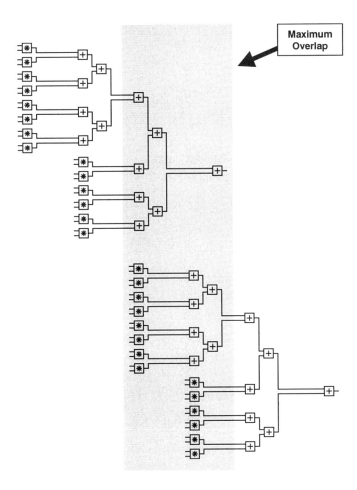

Figure 15-10: Pipelined design with an initialization interval of 2

15.6 Summary

This case study examines the implementation of a relatively simple behavioral description.

In this case study, the design space was first bounded by analyzing the "fastest" and "smallest" designs. Assuming that the throughput of even the fastest implementation was not acceptable, the design was pipelined, first using an initialization interval of 1 and then an initialization interval of 2. The first pipelined design resulted in a design with 3 times the throughput but at the cost of only a marginal area increase. The second pipelined design was significantly smaller, since data was being read only every other clock cycle.

This example highlights the many different ways that even a simple design can be implemented using behavioral synthesis. The four implementations that were considered in this chapter all have very different architectures, but can all be implemented in a mater of minutes.

Chapter 16

Case Study: Viterbi Decoding

16.1 Introduction

This chapter discusses the implementation of a Viterbi decoding algorithm. The Viterbi decoding algorithm is used to decode convolution codes and is found in many systems that receive digital data that might contain errors.

The chapter first discusses the details of the algorithm itself. It then discusses how the algorithm can be coded in a behavioral style and the synthesis issues that should be considered in order to produce a quality result.

16.2 The Algorithm

The Viterbi algorithm was first described in 1967 as a method for efficiently decoding convolution codes. Convolution codes are used to add redundancy to a stream of data prior to the transmission of that data through a potentially noisy channel.

At the receiver, the stream of data (which may now contain errors) is passed through a Viterbi decoder, which attempts to extract the *most likely* sequence of transmitted data. The structure of such a system is shown in Figure 16-1.

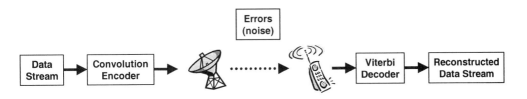

Figure 16-1: System that uses convolution codes and Viterbi decoding

16.2.1 Convolution Codes

Convolution codes are used to add redundancy to a stream of data. The addition of redundancy allows for the detection and (possible) correction of incorrectly transmitted data.

Unlike Hamming or Reed-Solomon codes that add redundancy to fixed-sized blocks of data, convolution codes add redundancy continuously – the output of a convolution encoder is dependent upon the current data *and* previously-transmitted data.

A convolution code is characterized by its rate, constraint length, and generator polynomials.

The rate of the convolution code is the ratio of the size of the input stream to the size of the output stream. The convolution code used in this example has a rate of ½, which means that each input bit produces 2 output bits.

The constraint length indicates the number of previous input data that must be examined (along with the current input data) to determine the output data. The convolution used in this example has a constraint length of 3.

The generator polynomials describe how past and current input data are used to determine the output data. The convolution code used in this example uses the following generator polynomials:

$out_{low} = in_t \wedge in_{t-2}$

$out_{high} = in_t \wedge in_{t-1} \wedge in_{t-2}$

These polynomials can be represented as a state machine, as shown in Figure 7-2.

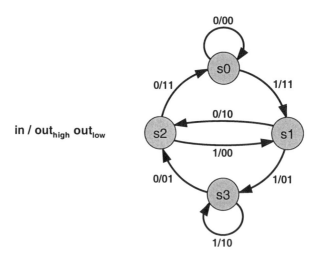

Figure 16-2: Implementation of a convolution encoder

Case Study: Viterbi Decoding

Figure 16-3 shows the convolution codes generated from a sample stream of data.

Input Stream	0	0	1	0	1	1	1
Next State	s0	s0	s1	s2	s1	s3	s3
Output Stream	00	00	11	10	00	01	10

Figure 16-3: Convolution codes generated from sample data stream

16.2.2 Trellis Diagram

A trellis diagram can be used to represent the operation of a state machine over time. In a trellis diagram, the states are listed vertically; time advances from left to right. Figure 16-4 shows the same input data of Figure 16-3, represented in a trellis diagram.

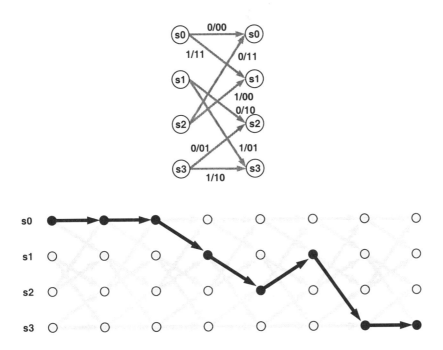

Figure 16-4: Trellis diagram showing sample data stream

Note that upper branches are followed when the input is '1'; lower branches are followed when the input is '0'.

16.2.3 Viterbi Decoder

A Viterbi decoder attempts to reconstruct a path through a trellis diagram based on a (potentially corrupted) stream of data. This is accomplished by selecting the "most likely" path through the trellis. In the decoding process, legal transitions are considered much more likely than illegal transitions. Not all transitions between state are legal. For example, there is no legal transition from state **s0** to state **s2**.

When new input data arrives, a probability value is calculated for every path between two sets of states. The probability is determined by calculating the *distance* between states, which is the number of bits that would have to be incorrect for the path to be taken. Consider the state **s0**. There are two possible paths to this state: one from state **s0** and one from state **s2**. These paths are highlighted in Figure 16-5.

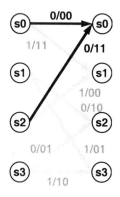

Figure 16-5: Possible transitions into state s0

If the input pattern is "00", the distance for the upper branch is 0, since the pattern associated with this branch is the same as the input pattern. The distance associated with the lower branch is 2, since the pattern associated with the branch differs from the input pattern for both bits. In this manner, distances are calculated for all 8 possible branches.

These distances are added to a set of *global* distances that are maintained for each state. These values represent the likelihood of being in a particular state at the time the new input data arrives.

The global distances are added to the branch distances in order to select, for each state, the most likely branch that would reach that state. The most likely branch is the branch that has the lowest distance – a larger the distance number indicates that a greater number of errors would have to occur for the path to be the correct one. This process is best illustrated by an example.

Case Study: Viterbi Decoding

Given a set of current global distances and an input pattern, Figure 16-6 shows the 8 branch distances, the 8 total distances, and the 4 selected branches.

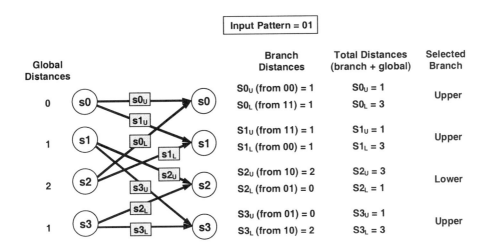

Figure 16-6: Branch selection based on global and branch distances

Once total distances have been calculated and branches have been selected, the branch selections are stored in a *survivor memory*. The survivor memory stores the most likely branches for some number of previous input data. The number of previous branch selections stored in the survivor memory is typically 4 to 5 times the constraint length. This is usually sufficient to adequately correct errors.

After the branches have been selected, the survivor memory is *backtracked* to determine the most likely error free input value. For each state, the survivor memory is used to determine the most likely previous state. Since the survivor memory indicates the most likely branch that was used to *get* to each state, it similarly indicates the most likely *previous* state. The Viterbi algorithm asserts that the process of backtracking will tend to correct input sequences that contain errors because the paths tend to converge as the survivor memory is traversed.

The backtracking process continues until the beginning of the survivor memory is reached. Assuming the paths have all converged into a single path, the direction of the last branch determines the decoded value. If the last branch was an *upper* branch, the output is '0', otherwise the output is '1'. If the paths have not converged, then some other decision-making process must be used to select the output value.

The backtracking process is best illustrated by an example. For simplicity, assume a survivor window depth of 7. There are four value stored for each entry in the survivor window. These values represent the most likely path (upper or lower) that was used to reach that state. Backtracking begins at the end of the survivor window and uses this information to trace backwards. Figure 16-7 shows two representations of a trellis. The top diagram shows the contents of the survivor memory, superimposed on the trellis. The bottom diagram shows the backtracked paths and how those paths tend to converge.

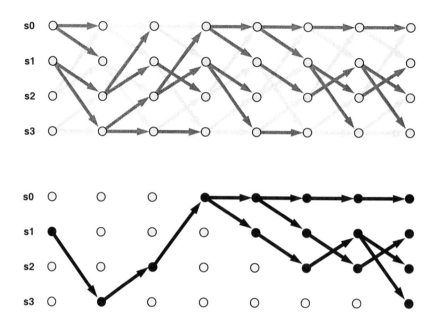

Figure 16-7: Backtracking through the survivor window

The figure shows that all of the paths have converged by the fourth step in backtracking. The final branch represents a lower path, so the output would be a '1'.

Note that the backtracking process introduces a delay between the input data stream and the output data stream. This delay is equal to the depth of the survivor memory. In the previous example, the decoded output is delayed by 7 data samples.

After backtracking, the global distance values are updated to reflect the new total distances to be use for the next input data. Since the global distances represent a running total that steadily increases as data arrives, the values must be periodically normalized to avoid overflow. This is accomplished by determining the smallest global distance and subtracting that value from all of the global distances. For example, if the global distances are 1, 2, 3, and 2, they would become 0, 1, 2, and 1 after normalization. Normalization maintains the relative difference between the values, but avoids overflow.

Finally, the contents of the survivor memory is shifted in preparation for the next input data.

16.3 Behavioral Description

While the Viterbi algorithm itself is somewhat difficult to understand, the coding of the behavioral description is quite straightforward. This design illustrates how coding at the behavioral level enhances the understandability of an algorithm. The flow through the algorithm can be easily followed. This example is quite small -- the entire description, with comments, is only about 200 lines.

Case Study: Viterbi Decoding

Coding at the behavioral level also simplifies support of the design and encourages its reuse. Consider modifying the behavioral description to utilize a different set of convolution codes, perhaps a code that generates 3 bits of output for each input bit. The trellis would have 8 possible states, instead of 4. Even though this is a radical change to the design, the overall flow of the algorithm is unchanged. The high-level nature of the behavioral description allows such changes to be made easily.

In this particular implementation, backtracking is not performed for every state. Rather, the backtracking process starts with most likely state, and traces a single path. This simplification is based on the assumption of the algorithm, that path tend to converge during the backtracking process.

The entire source code for the design can be found in Appendix C as well as on the CD that accompanies this book.

16.4 The Environment

The design was developed with a test bench. The test bench generates a random stream of data and perform convolution encoding on the data. Random errors are then inserted into the two data streams that are generated by convolution encoding. This represents the errors that might occur during transmission of the data through a wire or through the airwaves. The data steam, with errors, is processed by the Viterbi decoder. Finally, the output of the decoder is compared with the original signal, which has been delayed by the depth of the survivor window.

The overall structure of the test bench is shown in Figure 16-8.

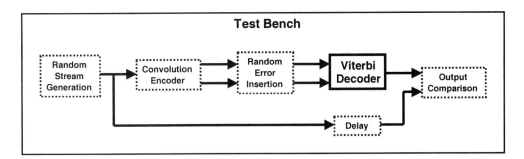

Figure 16-8: The design environment

The test bench does not have any inputs or outputs, but it does use generics to control its operation. There are generics to control the clock period, error rate, latency of input (i.e. how often input data arrives), as well as seeds for the random number generator. The test bench uses the random number generation function called **unique**, which is found in the *math_real* package.

The test environment also includes a TCL script to display the trellis with survivor window and backtracking information as well as the performance statistics of the decoder while the simulation is running.

A screen shot of a simulation run is shown in Figure 16-9.

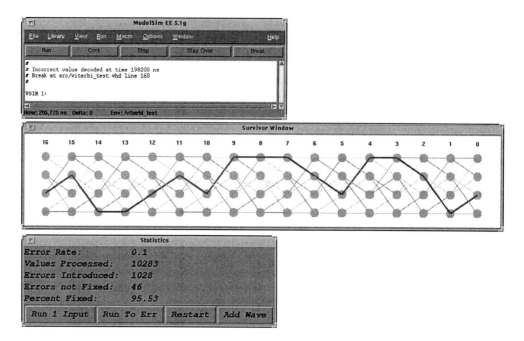

Figure 16-9: Sample simulation

16.5 Decoding Results

As can be seen in the statistics window of the sample simulation, the Viterbi decoder does an adequate job at correcting errors that were introduced into the data stream. The error rate was set to 10%, so for 10,283 data points, 1,028 errors were introduced. Of those errors, all but 46 were fixed. This means that 95.5% of the errors that were introduced could be fixed by Viterbi decoding. Of course, different error rates will produce different results.

Case Study: Viterbi Decoding

16.6 Behavioral Synthesis

16.6.1 Bounding the Design Space

As was done for the other case studies, consider the "fastest" and "smallest" solutions. Both of these solutions were generated using a small clock period, on the assumption that the design would be processing high-speed data.

This design contains only a small number of operations to schedule. The primary components are adders (that are used to accumulate the branch distances), comparators (that are used to perform branch selection and to determine the smallest branch distance), and subtractors (that are used to normalize the global branch distances).

As with the other case studies, the architectures that are discussed in this chapter were generated using Monet®, the behavioral design environment from Mentor Graphics Corporation. The Gantt chart for the "fastest" implementation is shown in Figure 16-10. Some of the operations were filtered from the Gantt chart display to simplify its appearance.

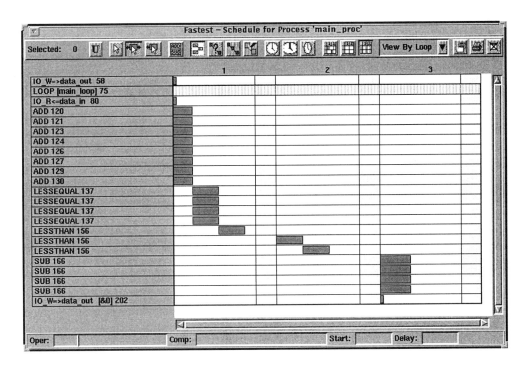

Figure 16-10: Gantt chart for "fastest" solution

The "smallest" implementation requires only one adder, one comparator, and one subtractor. This extends the schedule from 3 c-steps to 14 c-steps. The Gantt chart for the "smallest" implementation is shown in Figure 16-11.

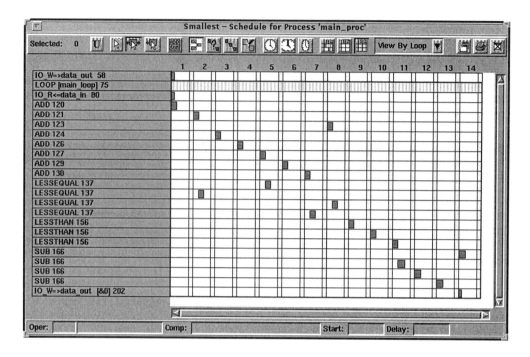

Figure 16-11: Gantt chart for "smallest" solution

Even though the "smallest" implementation uses the fewest data path elements, its final implementation in not smaller than the "fastest" implementation. In fact, the final implementation of the design with the smallest data path is approximately 1.5 times the size of the design with the larger data path. This difference is caused by the insertion of multiplexers and registers.

When fewer data path elements are used, multiplexers must be added to share the components between multiple c-steps. In addition, some values that are calculated early in the schedule must be held in registers until the data path elements become available in later c-steps.

> *The introduction of multiplexers and registers into a design as the result of resource sharing can significantly increase the size of some designs. The effect is particularly pronounced when the size of the data path elements that are being shared is relatively small when compared to the area of an equivalently sized multiplexer or register.*

As with the FIR filter example, the input data rate will determine what solutions are viable implementations of this design.

Consider the "fastest" solution. The design is scheduled in 3 c-steps. This means that the input data is only sampled every 3 clock cycles. If the input data arrives at a faster rate, much of the data will be lost, resulting in incorrect output.

Case Study: Viterbi Decoding

16.6.2 Pipelining the Design

Pipelining can be used to increase the throughput of the design, as was illustrated in the FIR filter case study. If the data arrives more frequently than every 3 clock periods (the latency of the "fastest" solution), then pipelining is required. Consider pipelining the design with an initialization interval of 1.

Recall from the discussion in Chapter 8 how the filter loop can be pipelined. First, the loop is duplicated as many times as necessary to determine the maximum overlap. This is shown in Figure 16-12. The figure does not show the individual operations, but rather the major steps that are performed in the Viterbi algorithm.

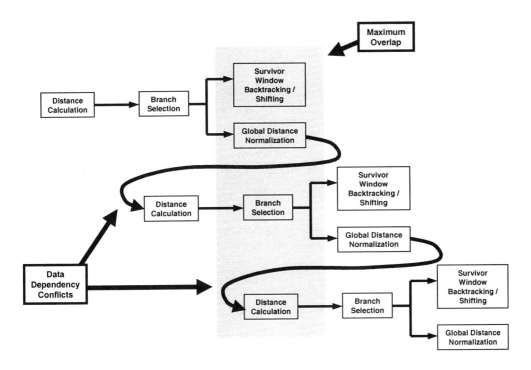

Figure 16-12: Attempting to pipeline the design with an initialization interval of 1

The figure shows that this design cannot be pipelined. Recall the steps that are performed in the algorithm. The first step is branch distance calculation. This step sums the distances for each branch with its corresponding global distance. One of the last steps involves updating the values of the global distances. This means that the processing of new input data cannot begin until the global distances from the previous calculation have been updated. This data dependency introduces a conflict when attempting to pipeline the design.

16.7 Summary

This case study examines the design and implementation of a Viterbi decoder. Although the algorithm is somewhat complex, the behavioral description is quite simple. The description highlights some of the benefits of describing designs at the behavioral level – the flow of the algorithm is easy to understand and could be easily modified for more complex convolution codes.

When analyzing the results of behavioral synthesis, both the "fastest" and "smallest" implementations were considered. Because the size of the data path elements in the design was small, the final implementation "fastest" design was actually smaller than the implementation that had the smallest data path. The size of the multiplexers and registers that were needed when the resources were shared was significant when compared to the size of the data path elements themselves.

Finally, a pipelined implementation of the design was considered. However, the data dependencies of the algorithm precluded the implementation of any pipelined architectures for the design.

This example highlights some of the issues that must be consider when using behavioral synthesis: first, the cost of sharing resources may be higher than simply allowing a greater number to be used; and second, that it may not be possible to pipeline a design when there are data dependencies that flow from one iteration of a loop to the next.

Appendix A

JPEG Source Code

jpeg_package.vhd

```vhdl
--------------------------------------------------------------------------------
--
-- Title: Package for JPEG Type and Constant definitions
--
-- Copyright (c) 1998,1999 by Mentor Graphics Corporation.  All rights reserved.
--
-- This source file may be used and distributed without restriction provided
-- that this copyright statement is not removed from the file and that any
-- derivative work contains this copyright notice.
--
--------------------------------------------------------------------------------

LIBRARY ieee;
USE ieee.std_logic_1164.ALL;
USE ieee.std_logic_arith.ALL;

PACKAGE jpeg_package IS

    SUBTYPE pixel IS signed( 8 DOWNTO 0 );
    TYPE pixel_matrix IS ARRAY ( integer RANGE <> ) OF pixel;

    SUBTYPE dct_entry IS signed( 10 DOWNTO 0 );
    TYPE dct_matrix IS ARRAY ( integer RANGE <> ) OF dct_entry;

    SUBTYPE temp_entry IS signed( 20 DOWNTO 0 );
    TYPE temp_matrix IS ARRAY ( integer RANGE <> ) OF temp_entry;

    SUBTYPE fraction IS signed( 9 DOWNTO 0 );
    TYPE fraction_matrix IS ARRAY ( integer RANGE <> ) OF fraction;

    --
    --              1 / sqrt( 8 )                              for i = 0
    -- c(i,j) =
    --              ( 1/2 ) * cos( ((2*j+1) * i * pi) / 16 )   for i > 0
    --
    --
    --    0.3535533906   0.3535533906   0.3535533906   0.3535533906
    --    0.3535533906   0.3535533906   0.3535533906   0.3535533906
    --    0.4903926402   0.4157348062   0.2777851165   0.0975451610
    --   -0.0975451610  -0.2777851165  -0.4157348062  -0.4903926402
    --    0.4619397663   0.1913417162  -0.1913417162  -0.4619397663
```

```
--  -0.4619397663 -0.1913417162  0.1913417162  0.4619397663
--   0.4157348062 -0.0975451610 -0.4903926402 -0.2777851165
--   0.2777851165  0.4903926402  0.0975451610 -0.4157348062
--   0.3535533906 -0.3535533906 -0.3535533906  0.3535533906
--   0.3535533906 -0.3535533906 -0.3535533906  0.3535533906
--   0.2777851165 -0.4903926402  0.0975451610  0.4157348062
--  -0.4157348062 -0.0975451610  0.4903926402 -0.2777851165
--   0.1913417162 -0.4619397663  0.4619397663 -0.1913417162
--  -0.1913417162  0.4619397663 -0.4619397663  0.1913417162
--   0.0975451610 -0.2777851165  0.4157348062 -0.4903926402
--   0.4903926402 -0.4157348062  0.2777851165 -0.0975451610
--

CONSTANT c : fraction_matrix( 0 TO 63 ) :=
  ( "0101101010", "0101101010", "0101101010", "0101101010",
    "0101101010", "0101101010", "0101101010", "0101101010",
    "0111110110", "0110101001", "0100011100", "0001100011",
    "1110011101", "1011100100", "1001010111", "1000001010",
    "0111011001", "0011000011", "1100111101", "1000100111",
    "1000100111", "1100111101", "0011000011", "0111011001",
    "0110101001", "1110011101", "1000001010", "1011100100",
    "0100011100", "0111110110", "0001100011", "1001010111",
    "0101101010", "1010010110", "1010010110", "0101101010",
    "0101101010", "1010010110", "1010010110", "0101101010",
    "0100011100", "1000001010", "0001100011", "0110101001",
    "1001010111", "1110011101", "0111110110", "1011100100",
    "0011000011", "1000100111", "0111011001", "1100111101",
    "1100111101", "0111011001", "1000100111", "0011000011",
    "0001100011", "1011100100", "0110101001", "1000001010",
    "0111110110", "1001010111", "0100011100", "1110011101"
  );

SUBTYPE zigzag_index IS integer RANGE 0 TO 63;
TYPE zigzag_index_matrix IS ARRAY ( 0 TO 63 ) OF zigzag_index;

--
--     0 - 1   2 - 3   4 - 5   6 - 7
--       /   /   /   /   /   /   /
--     8   9  10  11  12  13  14  15
--     | /   /   /   /   /   /   / |
--    16  17  18  19  20  21  22  23
--       /   /   /   /   /   /   /
--    24  25  26  27  28  29  30  31
--     | /   /   /   /   /   /   / |
--    32  33  34  35  36  37  38  39
--       /   /   /   /   /   /   /
--    40  41  42  43  44  45  46  47
--     | /   /   /   /   /   /   / |
--    48  49  50  51  52  53  54  55
--       /   /   /   /   /   /   /
--    56 -57  58 -59  60 -61  62 -63
--
CONSTANT c_zigzag_matrix : zigzag_index_matrix :=
  (   0,  1,  5,  6, 14, 15, 27, 28,
      2,  4,  7, 13, 16, 26, 29, 42,
      3,  8, 12, 17, 25, 30, 41, 43,
      9, 11, 18, 24, 31, 40, 44, 53,
     10, 19, 23, 32, 39, 45, 52, 54,
     20, 22, 33, 38, 46, 51, 55, 60,
     21, 34, 37, 47, 50, 56, 59, 61,
     35, 36, 48, 49, 57, 58, 62, 63
  );
```

Appendix A: JPEG Source Code

```vhdl
    CONSTANT zigzag_matrix : zigzag_index_matrix :=
        (   0,
            1,  8,
           16,  9,  2,
            3, 10, 17, 24,
           32, 25, 18, 11,  4,
            5, 12, 19, 26, 33, 40,
           48, 41, 34, 27, 20, 13,  6,
            7, 14, 21, 28, 35, 42, 49, 56,
           57, 50, 43, 36, 29, 22, 15,
           23, 30, 37, 44, 51, 58,
           59, 52, 45, 38, 31,
           39, 46, 53, 60,
           61, 54, 47,
           55, 62,
           63
        );

    FUNCTION divide(
        dividend    : signed( 10 DOWNTO 0 );
        divisor     : signed(  7 DOWNTO 0 )
    ) RETURN signed;

END jpeg_package;

PACKAGE BODY jpeg_package IS

    FUNCTION divide(
        dividend    : signed( 10 DOWNTO 0 );
        divisor     : signed(  7 DOWNTO 0 )
    ) RETURN signed IS
        VARIABLE v_dividend : unsigned( 18 DOWNTO 0 );
        VARIABLE v_divisor  : unsigned(  7 DOWNTO 0 );
        VARIABLE v_result   :   signed( 10 DOWNTO 0 );
        VARIABLE v_tmp      : unsigned(  8 DOWNTO 0 );
        VARIABLE sign       : std_logic;
    BEGIN
        sign := dividend( 10 ) XOR divisor( 7 );

        v_result   := ( OTHERS => '0' );
        v_divisor  := conv_unsigned( abs( divisor ), 8 );
        v_dividend := ( OTHERS => '0' );
        v_dividend( 10 DOWNTO 0 ) := conv_unsigned( abs( dividend ), 11 );

        FOR i IN 10 DOWNTO 0 LOOP

            v_tmp := v_dividend( 18 DOWNTO 10 );
            IF ( v_tmp >= v_divisor ) THEN
                v_result( i ) := '1';
                v_dividend := ( v_tmp - v_divisor ) & v_dividend( 9 DOWNTO 0 );
            END IF;

            v_dividend := v_dividend( 17 DOWNTO 0 ) & '0';

        END LOOP;

        IF ( sign = '1' ) THEN
            v_result := - v_result;
        END IF;

        RETURN v_result;
    END;
END jpeg_package;
```

jpeg_compress.vhd

```vhdl
--------------------------------------------------------------------------------
--
-- Title: JPEG Compression Algorithm
--
-- Copyright (c) 1998,1999 by Mentor Graphics Corporation.  All rights reserved.
--
-- This source file may be used and distributed without restriction provided
-- that this copyright statement is not removed from the file and that any
-- derivative work contains this copyright notice.
--
--------------------------------------------------------------------------------

LIBRARY ieee;
USE ieee.std_logic_1164.ALL;
USE ieee.std_logic_arith.ALL;
USE ieee.std_logic_signed.ALL;

USE work.jpeg_package.ALL;


ENTITY jpeg_compress IS
    PORT (
        clk             : IN  std_logic;
        rst             : IN  std_logic;
        load            : IN  std_logic;
        data_in         : IN  std_logic_vector( 7 DOWNTO 0 );
        quality         : IN  std_logic_vector( 3 DOWNTO 0 );
        ready_for_output : IN  std_logic;
        ready_for_input : OUT std_logic;
        output_valid    : OUT std_logic;
        data_out        : OUT std_logic_vector( 7 DOWNTO 0 )
    );
END jpeg_compress;


ARCHITECTURE behavioral OF jpeg_compress IS
BEGIN

    main_proc: PROCESS

        VARIABLE pixels : pixel_matrix( 0 TO 63 );  -- Input matrix storage (8x8)
        VARIABLE temp   : temp_matrix( 0 TO 63 );   -- Temp matrix storage  (8x8)
        VARIABLE dct    : dct_matrix( 0 TO 63 );    -- DCT matrix storage   (8x8)

        VARIABLE tmp          : signed( 20 DOWNTO 0 );
        VARIABLE quantum      : integer RANGE 0 TO 255;
        VARIABLE location     : integer RANGE 0 TO 63;

        SUBTYPE nibble IS std_logic_vector( 3 DOWNTO 0 );
        TYPE nibble_array IS ARRAY ( integer RANGE <> ) OF nibble;

        VARIABLE write_zeros    : boolean;
        VARIABLE zero_count     : integer RANGE 0 TO 8;
        VARIABLE dct_value      : signed( 30 DOWNTO 0 );
        VARIABLE dct_value_high : signed( 10 DOWNTO 0 );
        VARIABLE word           : dct_entry;
        VARIABLE word_abs       : dct_entry;
        VARIABLE nibbles        : nibble_array( 0 TO 7 );
        VARIABLE nibble_count   : integer RANGE 0 TO 7;
```

Appendix A: JPEG Source Code

```vhdl
    VARIABLE sub_word       : signed( 2 DOWNTO 0 );
    VARIABLE last_dc        : dct_entry;
    VARIABLE save_last_dc   : dct_entry;

    VARIABLE last_non_zero : integer RANGE 0 TO 63;
    VARIABLE eob           : std_logic;

    -- Define attributes (also in mgc_hls package)
    SUBTYPE resource IS integer;
    ATTRIBUTE map_to_module : string;
    ATTRIBUTE variables : string;

    -- Store the matrix data in memories
    CONSTANT pixel_ram                    : resource := 0;
    ATTRIBUTE variables OF pixel_ram      : CONSTANT IS "pixels";
    ATTRIBUTE map_to_module OF pixel_ram  : CONSTANT IS "ram_s_nRW";

    CONSTANT temp_ram                     : resource := 1;
    ATTRIBUTE variables OF temp_ram       : CONSTANT IS "temp";
    ATTRIBUTE map_to_module OF temp_ram   : CONSTANT IS "ram_s_nRW";

    CONSTANT dct_ram                      : resource := 2;
    ATTRIBUTE variables OF dct_ram        : CONSTANT IS "dct";
    ATTRIBUTE map_to_module OF dct_ram    : CONSTANT IS "ram_s_nRW";

BEGIN

    reset_loop: LOOP

        ready_for_input <= '0';
        output_valid <= '0';
        data_out <= ( OTHERS => '0' );
        last_dc := ( OTHERS => '0' );

        main_loop: LOOP

            read_loop:
            FOR i IN 0 TO 63 LOOP  -- pragma dont_unroll
                --
                -- Wait until next clock cycle to check for load signal.
                --
                WAIT UNTIL clk'EVENT AND ( clk = '1' );
                EXIT reset_loop WHEN ( rst = '1' );

                wait_load: WHILE ( load /= '1' ) LOOP

                    ready_for_input <= '1';

                    WAIT UNTIL clk'EVENT AND ( clk = '1' );
                    EXIT reset_loop WHEN ( rst = '1' );

                END LOOP;

                ready_for_input <= '0';

                pixels( i ) := conv_signed( conv_integer( '0' & data_in ) -
                                128, 9 );
            END LOOP;

            --
            -- Perform first matrix multiplication.
            --
            --    temp = pixels * ct
```

```vhdl
--
mult1:
FOR i IN 0 TO 7 LOOP   -- pragma dont_unroll
    FOR j IN 0 TO 7 LOOP   -- pragma dont_unroll
        tmp := ( OTHERS => '0' );
        FOR k IN 0 TO 7 LOOP
            tmp := tmp + pixels( i * 8 + k ) * c( j * 8 + k );
                                        -- ct( k * 8 + j )
        END LOOP;
        temp( i * 8 + j ) := tmp;
    END LOOP;
END LOOP;

--
-- Perform second matrix multiplication.
--
--     dct = c * temp
--
last_non_zero := 0;
mult2:
FOR i IN 0 TO 7 LOOP   -- pragma dont_unroll
    FOR j IN 0 TO 7 LOOP   -- pragma dont_unroll
        dct_value := ( OTHERS => '0' );
        FOR k IN 0 TO 7 LOOP
            dct_value := dct_value +
                    c( i * 8 + k ) * temp( k * 8 + j );
        END LOOP;

        --
        --     quantized = dct / quantum
        --
        quantum := 1 + ((1 + i + j) * conv_integer('0'& quality));
        dct_value_high := dct_value( 30 DOWNTO 20 );
        word := divide( dct_value_high, conv_signed(quantum, 8) );

        location := c_zigzag_matrix( i * 8 + j );
        IF ( ( word /= 0 ) AND ( location > last_non_zero ) ) THEN
            last_non_zero := location;
        END IF;

        dct( location ) := word;
    END LOOP;
END LOOP;

--
-- Quantize, compress and output results.
--
zero_count   := 0;
nibble_count := 0;

FOR i IN 0 TO 5 LOOP
    nibbles( i ) := "0000";
END LOOP;

eob := '0';

generate_results: FOR i IN 0 TO 63 LOOP   -- pragma dont_unroll

    word := dct( i );
    write_zeros := false;

    IF ( i = 0 ) THEN
        save_last_dc := last_dc;
```

```
            last_dc := word;
            word := word - save_last_dc;
        END IF;

        IF ( word = 0 ) THEN
            IF ( i /= 0 ) THEN
                zero_count := zero_count + 1;
                IF ( ( zero_count = 7 ) OR ( i = 63 ) ) THEN
                    write_zeros := true;
                END IF;
            END IF;
        ELSE
            IF ( zero_count /= 0 ) THEN
                write_zeros := true;
            END IF;
        END IF;

        IF ( write_zeros ) THEN
            nibbles( nibble_count ) :=
                        '0' & conv_std_logic_vector( zero_count, 3 );
            nibble_count := nibble_count + 1;
            zero_count := 0;
        END IF;

        IF ( ( word /= 0 ) OR ( i = 0 ) ) THEN
            word_abs := abs( word );
            IF ( word = 0 ) THEN
                nibbles( nibble_count ) := "0000";
                nibble_count := nibble_count + 1;
            ELSIF ( word_abs < 3 ) THEN    -- Range is -2 to 2
                sub_word := word( 2 DOWNTO 0 );
                IF ( sub_word = -1 ) THEN
                    sub_word := conv_signed( 0, 3 );
                ELSIF ( sub_word = -2 ) THEN
                    sub_word := conv_signed( -1, 3 );
                END IF;
                nibbles( nibble_count ) :=
                    "10" & std_logic_vector( sub_word( 1 DOWNTO 0 ) );
                nibble_count := nibble_count + 1;
            ELSIF ( word_abs < 16 ) THEN   -- Range is -15 to 15
                nibbles( nibble_count ) := "110" & word( 4 );
                nibble_count := nibble_count + 1;
                nibbles( nibble_count ) :=
                            std_logic_vector( word( 3 DOWNTO 0 ) );
                nibble_count := nibble_count + 1;
            ELSIF ( word_abs < 128 ) THEN  -- Range is -127 to 127
                nibbles( nibble_count ) := "1110";
                nibble_count := nibble_count + 1;
                nibbles( nibble_count ) :=
                            std_logic_vector( word( 7 DOWNTO 4 ) );
                nibble_count := nibble_count + 1;
                nibbles( nibble_count ) :=
                            std_logic_vector( word( 3 DOWNTO 0 ) );
                nibble_count := nibble_count + 1;
            ELSE   -- Range is -1023 to 1023
                nibbles( nibble_count ) := "1111";
                nibble_count := nibble_count + 1;
                nibbles( nibble_count ) :=
                        '0' & std_logic_vector( word( 10 DOWNTO 8 ) );
                nibble_count := nibble_count + 1;
                nibbles( nibble_count ) :=
                            std_logic_vector( word( 7 DOWNTO 4 ) );
                nibble_count := nibble_count + 1;
```

```vhdl
                        nibbles( nibble_count ) :=
                                  std_logic_vector( word( 3 DOWNTO 0 ) );
                        nibble_count := nibble_count + 1;
                    END IF;
                END IF;

                IF ( i = 63 ) THEN
                    nibbles( nibble_count ) := "0000";
                    nibble_count := nibble_count + 1;
                ELSIF ( i = last_non_zero ) THEN
                    nibbles( nibble_count ) := "0000";
                    nibble_count := nibble_count + 1;
                    nibbles( nibble_count ) := "0000";
                    nibble_count := nibble_count + 1;
                    eob := '1';
                END IF;

                nibble_loop: WHILE ( nibble_count > 1 ) LOOP

                    WAIT UNTIL clk'EVENT AND ( clk = '1' );
                    EXIT reset_loop WHEN ( rst = '1' );

                    out_rdy_loop: WHILE ( ready_for_output = '0' ) LOOP
                        WAIT UNTIL clk'EVENT AND ( clk = '1' );
                        EXIT reset_loop WHEN ( rst = '1' );
                    END LOOP;

                    data_out <= nibbles( 0 ) & nibbles( 1 );
                    output_valid <= '1';
                    WAIT UNTIL clk'EVENT AND ( clk = '1' );
                    EXIT reset_loop WHEN ( rst = '1' );

                    output_valid <= '0';
                    WAIT UNTIL clk'EVENT AND ( clk = '1' );
                    EXIT reset_loop WHEN ( rst = '1' );

                    nibbles( 0 ) := nibbles( 2 );
                    nibbles( 1 ) := nibbles( 3 );
                    nibbles( 2 ) := nibbles( 4 );
                    nibbles( 3 ) := nibbles( 5 );

                    nibble_count := nibble_count - 2;
                END LOOP;

                IF ( eob = '1' ) THEN
                    EXIT;
                END IF;
            END LOOP;

        END LOOP main_loop;

    END LOOP reset_loop;

    END PROCESS;

END behavioral;
```

Appendix B

FIR Filter Source Code

fir_filter.vhd

```vhdl
--------------------------------------------------------------------------------
--
-- Title: Finite Impulse Response (FIR) Filter
--
-- Copyright (c) 1998,1999 by Mentor Graphics Corporation.  All rights reserved.
--
-- This source file may be used and distributed without restriction provided
-- that this copyright statement is not removed from the file and that any
-- derivative work contains this copyright notice.
--
--------------------------------------------------------------------------------
LIBRARY ieee;
USE ieee.std_logic_1164.ALL;
USE ieee.std_logic_arith.ALL;
USE ieee.std_logic_signed.ALL;

ENTITY fir_filter IS
    PORT (
        clk             : IN  std_logic;
        rst             : IN  std_logic;
        load            : IN  std_logic;
        data_in         : IN  signed( 11 DOWNTO 0 );
        coeff_addr      : IN  unsigned( 3 DOWNTO 0 );
        start           : IN  std_logic;
        data_out        : OUT signed( 11 DOWNTO 0 )
    );
END fir_filter;

ARCHITECTURE behavioral OF fir_filter IS
BEGIN

    main_proc: PROCESS

        SUBTYPE  coeff_element IS signed( 8 DOWNTO 0 );
        TYPE     coeff_type    IS ARRAY ( integer RANGE <> ) OF coeff_element;
        VARIABLE coeff : coeff_type( 0 TO 14 );

        SUBTYPE  history_element IS signed( 11 DOWNTO 0 );
```

```vhdl
        TYPE     history_type    IS ARRAY ( integer RANGE <> ) OF history_element;
        VARIABLE history : history_type( 0 TO 14 );

        VARIABLE  sum : signed( 23 DOWNTO 0 );
        ATTRIBUTE unroll_new_instance : boolean;
        ATTRIBUTE unroll_new_instance OF sum : VARIABLE IS true;
    BEGIN

        reset_loop: LOOP

            data_out <= ( OTHERS => '0' );

            --
            -- Initialize all coefficients to zero
            --
            FOR i IN 0 TO 7 LOOP
                coeff( i ) := ( OTHERS => '0' );
            END LOOP;

            main_loop: LOOP

                config_loop: LOOP

                    --
                    -- Wait until next clock cycle to check for
                    -- load or filter signals.
                    --
                    WAIT UNTIL clk'EVENT AND ( clk = '1' );
                    EXIT reset_loop WHEN ( rst = '1' );

                    IF ( start = '1' ) THEN
                        EXIT config_loop;
                    END IF;

                    IF ( load = '1' ) THEN
                        coeff( conv_integer( '0' & coeff_addr ) ) :=
                                            signed( data_in( 8 DOWNTO 0 ) );
                    END IF;

                END LOOP config_loop;

                --
                -- Main filter loop
                --
                filter_loop: LOOP   -- pragma n_unroll 3
                                    -- pragma pipeline_init_interval 2

                    WAIT UNTIL clk'EVENT AND ( clk = '1' );
                    EXIT reset_loop WHEN ( rst = '1' );

                    history( 0 ) := signed( data_in );

                    sum := ( ( ( history(  0 ) * coeff(  0 ) +
                                 history(  1 ) * coeff(  1 ) ) +
                               ( history(  2 ) * coeff(  2 ) +
                                 history(  3 ) * coeff(  3 ) ) ) +
                             ( ( history(  4 ) * coeff(  4 ) +
                                 history(  5 ) * coeff(  5 ) ) +
                               ( history(  6 ) * coeff(  6 ) +
                                 history(  7 ) * coeff(  7 ) ) ) ) +
                           ( ( ( history(  8 ) * coeff(  8 ) +
                                 history(  9 ) * coeff(  9 ) ) +
```

```
                            ( history( 10 ) * coeff( 10 ) +
                              history( 11 ) * coeff( 11 ) ) ) +
                          ( ( history( 12 ) * coeff( 12 ) +
                              history( 13 ) * coeff( 13 ) ) +
                              history( 14 ) * coeff( 14 ) ) );

                data_out <= sum( 23 DOWNTO 12 );

                FOR i IN 14 DOWNTO 1 LOOP
                    history( i ) := history( i - 1 );
                END LOOP;

            END LOOP filter_loop;

        END LOOP main_loop;

    END LOOP reset_loop;

  END PROCESS;

END behavioral;
```

Appendix C

Viterbi Source Code

viterbi.vhd

```vhdl
--------------------------------------------------------------------------------
--
-- Title: Viterbi Decoder Algorithm
--
-- Copyright (c) 1998,1999 by Mentor Graphics Corporation.  All rights reserved.
--
-- This source file may be used and distributed without restriction provided
-- that this copyright statement is not removed from the file and that any
-- derivative work contains this copyright notice.
--
--------------------------------------------------------------------------------

LIBRARY ieee;
USE ieee.std_logic_1164.ALL;
USE ieee.std_logic_arith.ALL;
USE ieee.std_logic_signed.ALL;

ENTITY viterbi IS
    PORT (
        clk           : IN  std_logic;
        rst           : IN  std_logic;
        data_in       : IN  std_logic_vector( 1 DOWNTO 0 );
        data_out      : OUT std_logic
    );
END viterbi;


ARCHITECTURE behavioral OF viterbi IS

BEGIN

    main_proc: PROCESS

        CONSTANT window_length : integer := 16;

        SUBTYPE  survivor_elements IS std_logic_vector( 0 TO 3 );
        TYPE     survivor_window_type IS ARRAY ( integer RANGE <> ) OF
                                                           survivor_elements;
        VARIABLE survivor_window : survivor_window_type(window_length-1 DOWNTO 0);
        VARIABLE survivors                     : survivor_elements;
```

```vhdl
        VARIABLE backtrack_survivors  : survivor_elements;

        TYPE     distance_array_type IS ARRAY ( 0 TO 3 ) OF integer RANGE 0 TO 3;
        VARIABLE distance             : distance_array_type;
        VARIABLE global_distance      : distance_array_type;

        VARIABLE data_in_v            : std_logic_vector( 1 DOWNTO 0 );

        TYPE branch_distance_array_type IS ARRAY (0 TO 3) OF integer RANGE 0 TO 7;
        VARIABLE upper_branch_distance : branch_distance_array_type;
        VARIABLE lower_branch_distance : branch_distance_array_type;
        VARIABLE branch_distance       : branch_distance_array_type;
        VARIABLE minimum_branch        : integer RANGE 0 TO 7;

        SUBTYPE  state_type IS integer RANGE 0 TO 3;
        VARIABLE state                 : state_type;
        VARIABLE branch_direction      : std_logic;
BEGIN

        data_out <= '0';

        --
        -- Initialize survivor memory
        --
        FOR i IN window_length - 1 DOWNTO 0 LOOP
            survivor_window( i ) := ( OTHERS => '0' );
        END LOOP;

        --
        -- Initialize global distances
        --
        FOR i IN 3 DOWNTO 1 LOOP
            global_distance( i ) := 2;
        END LOOP;
        global_distance( 0 ) := 0;

        main_loop: LOOP

            WAIT UNTIL ( clk'EVENT AND clk = '1' );
            EXIT main_loop WHEN ( rst = '1' );

            data_in_v := data_in;

            --
            -- Calculate distances (# of bits which are different)
            --
            CASE data_in_v IS
                WHEN "00" =>
                    distance( 0 ) := 0;
                    distance( 1 ) := 1;
                    distance( 2 ) := 1;
                    distance( 3 ) := 2;
                WHEN "01" =>
                    distance( 0 ) := 1;
                    distance( 1 ) := 0;
                    distance( 2 ) := 2;
                    distance( 3 ) := 1;
                WHEN "10" =>
                    distance( 0 ) := 1;
                    distance( 1 ) := 2;
                    distance( 2 ) := 0;
                    distance( 3 ) := 1;
```

Appendix A: JPEG Source Code

```vhdl
            WHEN "11" =>
                distance( 0 ) := 2;
                distance( 1 ) := 1;
                distance( 2 ) := 1;
                distance( 3 ) := 0;
            WHEN OTHERS =>
                NULL;
        END CASE;

        --
        -- Add-Compare-Select (ACS)
        --
        acs_loop: FOR i IN 0 TO 3 LOOP

            --
            -- Calculate distances for the upper and lower branches
            --
            CASE i IS

                WHEN 0 => upper_branch_distance( i ) := distance( 0 ) +
                                                       global_distance( 0 );
                          --    st0 -- "00" (0) => st0
                          lower_branch_distance( i ) := distance( 3 ) +
                                                       global_distance( 2 );
                          --    st2 -- "11" (3) => st0

                WHEN 1 => upper_branch_distance( i ) := distance( 3 ) +
                                                       global_distance( 0 );
                          --    st0 -- "11" (3) => st1
                          lower_branch_distance( i ) := distance( 0 ) +
                                                       global_distance( 2 );
                          --    st2 -- "00" (0) => st1

                WHEN 2 => upper_branch_distance( i ) := distance( 1 ) +
                                                       global_distance( 1 );
                          --    st1 -- "01" (1) => st2
                          lower_branch_distance( i ) := distance( 2 ) +
                                                       global_distance( 3 );
                          --    st3 -- "10" (2) => st2

                WHEN 3 => upper_branch_distance( i ) := distance( 2 ) +
                                                       global_distance( 1 );
                          --    st1 -- "10" (2) => st3
                          lower_branch_distance( i ) := distance( 1 ) +
                                                       global_distance( 3 );
                          --    st3 -- "01" (1) => st3

            END CASE;

            --
            -- Select the surviving branch and fill appropriate value
            -- into the survivor window
            --
            IF ( upper_branch_distance(i) <= lower_branch_distance(i) ) THEN
                branch_distance( i ) := upper_branch_distance( i );
                survivors( i ) := '0';
            ELSE
                branch_distance( i ) := lower_branch_distance( i );
                survivors( i ) := '1';
            END IF;

        END LOOP;
```

```vhdl
        survivor_window( 0 ) := survivors;

        --
        -- Find the minimum branch distance and the ending state
        --
        minimum_branch := branch_distance( 0 );
        state := 0;

        find_minimum: FOR i IN 1 TO 3 LOOP
            IF ( branch_distance( i ) < minimum_branch ) THEN
                minimum_branch := branch_distance( i );
                state := i;
            END IF;
        END LOOP;

        --
        -- Subtract the minimum distance to avoid overflow
        --
        normalize: FOR i IN 0 TO 3 LOOP
            global_distance( i ) := branch_distance( i ) - minimum_branch;
        END LOOP;

        --
        -- Backtrack the survivor window from the most-likely state
        --
        backtrack: FOR i IN 0 TO window_length - 1 LOOP

            backtrack_survivors := survivor_window( i );
            branch_direction := backtrack_survivors( state );

            CASE state IS
                WHEN 0 | 1 => IF ( branch_direction = '0' ) THEN
                                  state := 0;
                              ELSE
                                  state := 2;
                              END IF;
                WHEN 2 | 3 => IF ( branch_direction = '0' ) THEN
                                  state := 1;
                              ELSE
                                  state := 3;
                              END IF;
            END CASE;
        END LOOP;

        --
        -- Shift the survivor window values
        --
        shift: FOR i IN window_length - 1 DOWNTO 1 LOOP
            survivor_window( i ) := survivor_window( i - 1 );
        END LOOP;

        --
        -- Generate output
        --
        data_out <= branch_direction;

        END LOOP main_loop;

    END PROCESS;

END behavioral;
```

Glossary

Abstract Data Type A data type that does not readily translate to hardware. A synthesis tool must translate abstract types into non-abstract types (such as *std_logic*).

ALAP Scheduling A scheduling strategy in which operations are scheduled in the latest possible clock cycle. ALAP stands for *As Late As Possible*.

Allocation See *Resource Allocation*

ASAP Scheduling A scheduling strategy in which operations are scheduled in the earliest possible clock cycle. ASAP stands for *As Soon As Possible*.

Attribute The mechanism in VHDL used to associate a property with an object. Attributes can be used to place user-specified synthesis controls directly in the VHDL source.

Behavioral Synthesis Writing a circuit description at the un-clocked algorithmic level, with few or no implementation details, then using tools to generate a clocked netlist of components. Behavioral synthesis tools take a behavioral description and automatically schedule the operations into clock cycles.

Binding See *Component Binding*

c-step See *Control Step*

CDFG See *Control / Data Flow Graph*

Chaining
The cascading of two or more data-dependent operations into the same clock cycle without the need to register intermediate results.

Component Binding
The relationship between a component and the operations in a CDFG is defined by the component bindings that are defined for that component. A component binding defines how a component can be used to implement the functionality of the operator.

Control / Data Flow Graph
A Control / Data Flow Graph (CDFG) is an extension of a DFG that allows the representation of control structures, such as conditionals and loops.

Control Step
The term c-step (or control step) refers to a state in the state machine that is generated during behavioral synthesis. For a loop structure, a Gantt chart shows the control steps (states) for scheduling the operations for a single iteration of the loop. Since a loop may execute multiple times, the actual number of clock cycles required to complete the loop may be significantly different.

Control Unit
The abstract representation of the control portion of the design.

Constraint
A user-specified synthesis control expressed as a value or a set of values for a particular parameter, such as maximum area or delay.

Cycle-Accurate Model
A model generated after scheduling, but before component binding. The cycle-to-cycle behavior of the netlist is identical to the final design, however, operators are netlisted in an abstract manner (for example, +, instead of a collection of gates) to still facilitate extremely fast simulation.

Cycle-Fixed I/O Mode
A mode of scheduling where there is an exact cycle-by-cycle match with the pre-synthesis simulation. In this mode, scheduling cannot add, delete, or move clock edges that are defined in the HDL source.

Glossary

Data Flow Graph
A Data Flow Graph (DFG) depicts the inputs and outputs of a design, the operations used in the design, and the data dependencies that define the flow of data from the inputs to the outputs.

Data Path
The portion of the design that computes data, from primary inputs or from intermediate input. Typically, the data path is controlled by the control unit.

Dependency Arc
A line between operations in a DFG or CDFG that indicates the order in which the operations must be performed. The operation which starts the arc must be complete before the operation that ends the arc can begin.

DFG
See *Data Flow Graph*

Fixed I/O Mode
See *Cycle-Fixed I/O Mode*

Fixed Mode
See *Cycle-Fixed I/O Mode*

Force-Directed Scheduling
Force-directed scheduling is a scheduling algorithm that can be used to address the time-constrained scheduling problem. The algorithm determines the "force" (i.e. area cost) associated with scheduling each operation into a particular clock cycle within its mobility window. The algorithm tends to produce a uniform distribution of operations across the schedule.

Free Mode
See *Free-Floating I/O Mode*

Free-Floating I/O Mode
Free-Floating I/O mode is the least restrictive scheduling mode used by behavioral synthesis tools. In free mode, synthesis is totally free to add or delete clock cycles and to move I/O operations. This provides the tool with maximum freedom to explore architectural alternatives. However, since free I/O mode may move I/O operations, the I/O protocol of the scheduled design may not match the I/O protocol of the behavioral description.

Free-Floating Mode	See *Free-Floating I/O Mode*
Gantt Chart	A Gantt chart, often used to display work schedules, shows the actual time required to perform each operation and the data dependencies between operations.
Handshaking	The introduction of additional signals to indicate when data should be passed between models. This allows the models to independently process data and only synchronize when it is necessary to pass data back and forth.
HDL	Short for Hardware Description Language, it refers to a textual representation of a circuit, usually VHDL or Verilog®. It is also sometimes used as a generic term referring to the Verilog® language.
High-Level Synthesis	See *Behavioral Synthesis*
ILP	See *Integer Linear Programming*
Initialization Interval	The initialization interval of a pipelined component or loop specifies the rate at which the component or loop can begin processing new input data. Most pipelined components have an initialization interval of 1, which means that the component can begin processing new input data every clock cycle.
In-lined Function	When the contents of a function or procedure is in-lined, the function or procedure call is replaced with the contents of the subprogram. The operations inside of the subprogram are scheduled along with every other operation in the design.

Glossary

Integer Linear Programming
A mathematical technique that can be used to address the scheduling problem. Unlike algorithms that are based on heuristics, integer linear programming can provide optimal solutions to the time and resource-constrained scheduling problems. Unfortunately, these techniques are so computationally intensive, they are usually unusable for any real-life scheduling problems.

Lifetime Analysis
Lifetime analysis is used to determine the clock cycles in which data is valid. This information can be used to share registers in the implementation of a scheduled design.

List Scheduling
List scheduling is a scheduling algorithm that can be used to schedule a resource-constrained design. The algorithm selects operations to be scheduled from a *ready list*, which is a prioritized list of every operation that could be scheduled in the current clock cycle without violating data dependencies.

Loop Pipelining
Loop pipelining is a feature of behavioral synthesis tools that allows a loop to be implemented with the structure of a pipelined component. This technique can be used to increase the throughput of a loop, but often at the expense of additional component area.

Loop Unrolling
Loop unrolling is a transformation that can be applied to a loop construct. When a loop is unrolled, the statements inside the loop are copied as many times as the loop is being unrolled.

Memory Inferencing
The mapping of an array in a behavioral description to a memory device. Using arrays as "memory devices" provides memory technology independence as well as easier design development and debugging.

Mobility
The mobility of an operation is defined by the range of c-steps in which the operation could be scheduled without changing the overall schedule length.

Multi-cycle Operation	An operation that is scheduled across two or more clock periods is a multi-cycle operation. If a multi-cycle operation is combinational, its inputs must be held stable while the output is computed. If the operation is pipelined, a new calculation can begin prior to the completion of the current calculation.
Operation	An instance of an operator in a design. In any design, there may be many operations that represent the functionality of a single operator.
Operator	An abstract representation of design functionality, such as ADD (for addition) or MUL (for multiplication). Operators are extracted from a behavioral description when constructing a CDFG.
Pragma	A user-specified synthesis tool directive that is placed directly in the VHDL source as a comment. Another tool, such as a VHDL simulator, will analyze this simply as a comment and ignore it.
Preserved Function	When a function is preserved, the entire function is scheduled as a single combinational operator. This means that multiple uses of the function can be shared, if appropriate. However, the operations inside the function can not be shared with other similar operations in the body of the design.
Resource Allocation	The assignment of hardware component types to data operations.
Resource Sharing	The general term that refers to the optimization of a design by allowing the sharing of single resources by multiple operations.
RTL	Register-Transfer Level, a circuit description style in which all clock edges, storage elements, and transformations of values are manifest in the source.
RTL Synthesis	The process by which an RTL description is transformed into a gate-level, technology-specific netlist.

Glossary

Scheduling
The assignment of data operations to clock cycles such that constraints and data flow relationships are met.

Signal Assignments
In VHDL, an assignment to a signal. In Cycle-Fixed I/O mode, such an assignment is fixed to a particular clock cycle. In Superstate-Fixed I/O, a signal assignment is fixed to a superstate.

State Encoding
The assignment of particular bit patterns to the abstract states of a state machine. The encoding depends on the encoding style. By default, most synthesis tools use binary encoding.

Structured RTL Netlist
The final netlist produced by behavioral synthesis, consisting of data path components, multiplexer, registers, and a state machine. The data path portion of the design is entirely structural. The state machine portion of the design can be represented as an RTL description.

Superstate
In superstate I/O mode, WAIT statements in the behavioral description define the boundaries of a *superstate*. A superstate can represent *one or more* clock cycles. I/O operations (signal reads and signal writes) can not cross a superstate boundary and if a superstate is represented by more than one c-step, output signals are only written to in the last c-step within the superstate.

Superstate Mode
See *Superstate-Fixed I/O Mode*

Superstate-Fixed I/O Mode	Unlike fixed I/O mode, in which the time between consecutive WAIT statements represents a *single* clock cycle, in superstate I/O mode, consecutive WAIT statements can represent *one or more* clock cycles. In superstate I/O mode, WAIT statements in the behavioral description define the boundaries of a *superstate*. I/O operations (signal reads and signal writes) can not cross a superstate boundary and if a superstate is represented by more than one c-step, output signals are only written to in the last c-step within the superstate. Superstate mode is best used when the sequence of I/O operations is important but the exact cycle-to-cycle correspondence between simulation of the behavioral description and the synthesized design is not.
Variable Assignments	In VHDL, an assignment to a variable. Variable assignments are not fixed and can be moved by the scheduler, independent of the I/O scheduling mode.
VHDL	VHSIC hardware description language, the textual representation language for circuits defined by IEEE standard 1076-1993.
Wrapper Model	If any of the ports of a behavioral design have data types that are modified during synthesis, a "wrapper" model can be created to allow the resultant design to communicate with the surrounding behavioral environment. The wrapper model preserves the original interface of the behavioral model and converts the data on these ports in the same manner as the synthesis tool.

References and Resources

J. Bhasker. *A VHDL Primer, Revised Edition*. Prentice-Hall, 1995.

Giovanni De Micheli. *Synthesis and Optimization of Digital Circuits*. McGraw-Hill, 1994.

G. David Forney, Jr. "The Viterbi Algorithm". *Proceedings of the IEEE, Volume 61, Number 3*. IEEE, 1973.

Daniel D. Gajski, Nikil D. Dutt, Allen C. H. Wu, and Steve Y. L. Lin. *High-Level Synthesis: Introduction to Chip and System Design*. Kluwer, 1992.

Daniel D. Gajski and Loganath Ramachandran. "Introduction to High-Level Synthesis". *IEEE Design and Test of Computers*. IEEE, 1994.

Emmanuel C. Ifeachor and Barrie W. Jervis. *Digital Signal Processing: A Practical Approach*. Addison-Wesley, 1993.

David W. Knapp. *Behavioral Synthesis: Digital System Design Using the Synopsys® Behavioral Compiler™*. Prentice-Hall, 1996.

Mark Nelson and Jean-Loup Gailly. *The Data Compression Handbook, Second Edition*. M&T Books, 1996.

Martin de Prycker. *Asynchronous Transfer Mode: Solutions for Broadband ISDN, Third Edition*. Prentice-Hall, 1995.

Douglas J. Smith. *HDL Chip Design: A Practical Guide for Designing, Synthesizing, and Simulating ASICs and FPGAs Using VHDL or Verilog*. Doone Publications, 1996.

Robert A. Walker and Samit Chaudhuri. "Introduction to the Scheduling Problem". *IEEE Design and Test of Computers*. IEEE, 1995.

CD-ROM

CD-ROM Contents

The CD-ROM contains the source code and test benches for the three case studies discussed in Chapters 14, 15, and 16. The files associated with these case studies can be found in the directories **jpeg**, **fir_filter**, and **viterbi**, respectively. Each of these directories contains a **README** file that explains the contents of the directory.

Specific comments or feedback regarding this book or its examples can be sent to the author at john_p_elliott@yahoo.com

CD-ROM License Agreement

By opening the CD-ROM package, you are agreeing to be bound by the following agreement:

The examples contained on the CD-ROM are copyrighted by Mentor Graphics Corporation.

THIS SOFTWARE IS PROVIDED "AS-IS" WITHOUT WARRANTY OF ANY KIND, NEITHER EXPRESS NOR IMPLIED, INCLUDING BUT NOT LIMITED TO THE IMPLIED WARRANTIES OF MERCHANTABILITY AND FITNESS FOR A PARTICULAR PURPOSE.

IN NO EVENT SHALL MENTOR GRAPHICS CORPORATION OR ITS LICENSORS, NOR THE PUBLISHER OR ITS DEALERS OR DISTRIBUTORS BE LIABLE FOR ANY CLAIMS AGAINST YOU BY ANY THIRD PARTY NOR SHALL MENTOR GRAPHICS CORPORATION OR ITS LICENSORS, NOR THE PUBLISHER OR ITS DEALERS OR DISTRIBUTORS BE LIABLE FOR ANY DAMAGES WHATSOEVER, INCLUDING DAMAGES FOR LOSS OF PROFITS, BUSINESS INTERRUPTION, LOSS OF BUSINESS INFORMATION, OR OTHER PECUNIARY LOSS OR FOR INDIRECT, SPECIAL, INCIDENTAL, OR OTHER CONSEQUENTIAL DAMAGES ARISING OUT OF THE USE OF OR INABILITY TO USE THE SOFTWARE, INCLUDING THE NON-FUNCTIONALITY OF THE SOFTWARE, EVEN IF MENTOR GRAPHICS HAS BEEN ADVISED OF THE POSSIBILITY OF SUCH DAMAGES.

Index

A

abstract type, *see data types*
access type, *see data types*
architecture-independent, 4
arrays, 50-54
  - mapping to memory, *see memory*
  - splitting, 51-52
  - non-constant indexing, 52-53
algorithm, development of, 10
allocation, *see resource allocation*
architectures (hardware), 5-7, 16, 105
  - exploring alternate, 11
  - resource allocation, 16
architecture, VHDL, 64-65
  - coding style, 214-216
as-late-as-possible (ALAP) scheduling, 30
ASSERT statement, 168-169, 210
as-soon-as-possible (ASAP) scheduling, 30
asynchronous memory, *see memory*
asynchronous reset, *see reset*
attributes
  - map_to_module, 157
  - variables, 158

B

behavioral description, 6, 80
  - for FIR filter, 264-265
  - for JPEG compression algorithm, 254
  - for Viterbi decoder, 278-279
behavioral design, 5, 6, 12-13
behavioral model, 12
behavioral synthesis, 10, 25, 39, 69
  - advantages of, 7
  - limitations of, 8
  - motivation for, 1-9
binding, *see component binding*
bit / bit_vector type, *see data types*
boolean type, *see data types*

C

CASE statement, 113
case studies
  - FIR filter, 263-265
  - JPEG compression, 245-247
  - Viterbi decoder, 274-276
chained operations, 22, 34
clock
  - coding style, 215-216
  - cycle, 6
  - edges, 69-70, 73
  - generation, 204
  - period, 6
clocked process, *see process*

coding style, for behavioral synthesis, 214
- architectures, 214-216
- clock edges, 215-216
- conditional statements, 228-230, 238-239
- data types, 218-220
- entities, 214-216
- fixed I/O mode, 225
- free I/O mode, 241-242
- loops, 232-234, 239-240
- processes, 214-216
- resets, 216-217
- scheduling modes, 225
- superstate I/O mode, 235-238

combinational components, 15
combinational process, *see process*
communication
- inter-process, 181
- with external models, 181

component binding, 27-29, 37-38, 174
- multiple possible, 34

component declaration, 203
component instantiation, 64, 203
concurrent assignment statement, 64
conditional statements
- coding style, 228-230, 238-239
- scheduling, 109, 124

constant declaration, 178
constrained designs, 30
control / data flow graph (CDFG), 26-27
control logic, 18
conversion functions, 45-47, 61

cycle-accurate simulation, 39-40, 250-251
cycle-fixed I/O mode, 106-109, 121, 199
- advantages and disadvantages, 117-118
- coding style, 225
- scheduling conditional branches in, 109
- scheduling loops in, 114
- synchronizing communication with, 182
- testing designs in, 109

D

data dependencies, 14, 27, 68
- memory operations, 167-171

data flow graph (DFG), 14-15, 17, 26, 68
data path elements, 5, 19
data path extraction, 39
data types, 41
- abstract, 42, 60
- access type, 54
- array type, 50-54, 224
- bit / bit_vector type, 42, 219
- boolean type, 43, 219
- coding style, 218-220
- enumerated type, 48, 60, 224
- floating point type, 54
- integer type, 47, 223
- interface, 199-200
- modification during synthesis, 58
- non-abstract, 41-42
- not supported for synthesis, 54

Index

- physical type, 54
- record type, 49-50, 224
- signed / unsigned, 41, 43, 219
- std_logic / std_logic_vector, 41, 43-44, 219

dependency arc, 26

design cycle, 200

design space exploration, 5

Design Compiler™, 3

discrete cosine transform (DCT), 265-267

E

entity, VHDL, 57, 202
 - coding style, 214-216

enumerated type, *see data types*

EXIT statements, 81-82
 - in pipelined loop, 149-150

F

FIR filter case study, 263-265

fixed I/O mode, *see cycle-fixed I/O mode*

free-floating I/O mode, 130-132
 - coding style, 241-242
 - synchronizing communication with, 182
 - testing designs in, 132

floating point type, *see data types*

force-directed scheduling, 32

FOR loop, 77, 91-92
 (see also loop)
 - execution time, 78
 - translation of, 91

- unrolling, 78

functions, 173-174
 - in-lining, 173
 - mapped to operator, 174
 - preserving, 176

G

Gantt chart, 20-21, 95-96

GENERIC, VHDL, 202

H

handshaking, 122, 129, 131, 181, 184
 - full, 184-185
 - input, 93, 97, 186-189
 - output, 190-191
 - partial, 184-185
 - scheduling issues, 191-192
 - superstate I/O mode, 192-194

hardware, generation of, 47, 50-51, 78

I

IEEE 1076.3 standard, 46

IF statement, 111

infinite loop, 77, 80, 88
 (see also loops)

initialization interval
 - of a pipelined component, 135-136, 146

initialization sequence, 85

in-lining, 173

input handshaking, *see handshaking*

integer linear programming (ILP), 34

integer type, *see data types*

Intellectual Property (IP) blocks, 176

inter-process communication, *see communication and handshaking*

I/O operations, 119

I/O scheduling modes, 105
- coding style, 225

I/O timing, 72, 117, 198

J

JPEG compression algorithm case study, 245-247

L

latency, 11, 71, 118
- of a pipelined component, 135-137, 146

lifetime analysis, 35-36

lifetimes, non-overlapping, 36

list scheduling, 31-33

logic synthesis, 5

loops, 77

(*see also FOR loop, infinite loop, and WHILE loop*)
- coding style, 232-234, 239-240
- complete unrolling, 98-99
- constructs, 79
- execution time, 96
- in CDFG, 27
- partial unrolling, 100-101
- pipelining, 142-143
- rolled, 154
- scheduling, 94
- unrolled, 154
- unrolling, 98-102

loop pipelining, *see loops*

M

memory, 18-19, 156-160
- addresses, 156-160
- aligned packing, 158
- asynchronous, 163-165
- bottlenecks, 159
- data dependencies, 167-171
- differing width variables, 159
- explicit packing, 158
- in RTL design, 156
- mapping from arrays, 156
- packing methods, 158
- ports, 166
- read and write operations, 161
- synchronous, 160-163
- timing constraints, 164

messages, in test benches, 210-211

mobility, 32

multi-cycle components, 140-141

multi-cycle operations, 22, 34

multi-cycle paths, 140-141

N

netlist, 39-40
- after scheduling, 39-40
- cycle-accurate, 39-40
- gate-level, 41
- structured RTL, 40

NEXT statements, 92-93, 114
- in pipelined loop, 149-150

Index

O

operations
- binding of, 175
- I/O, 119
- memory read and write, 161
- negligible delay, 116

P

packages, 173, 178
parallel computation, 139
physical type, *see data types*
pipelined components, 135-136, 138, 141
- implementation of, 135, 138
- initialization interval of, 135-136, 145
- latency, 135-137, 146
- minimum clock period of, 135-136
- multiplier, 139
- output delay of, 135-136

pipelined loops, 141
- data dependencies in, 151
- EXIT and NEXT statements in, 149-150
- implementation, 145, 148
- in FIR filter, 270-271
- in Viterbi decoder, 283
- restrictions on, 154
- scheduling, 143
- simulation of, 152
- stage in, 145

pipelined multiplier, 139
pipelining, *see pipelined components and pipelined loops*

ports, 57
- with abstract types, 58

procedure, VHDL, 173, 177

processes, 65-67
- clocked, 65, 82
- coding style, 214-216
- combinational, 65
- multiple, 70, 73-74

R

ready list, 31
records, 49
- splitting, 50
register allocation, 35
register sharing, 36
Register Transfer Level, *see RTL*
reset
- asynchronous, 87
- coding style, 216-217
- complex sequence, 85-86
- generation, 204
- simulation of, 83-85
- synchronous, 82

reset signal, 82
- using attribute to specify, 86

resource allocation, 16-17, 27, 29
- sufficient, 29

resource utilization, 147-148
RTL description, 5
RTL design, 5
RTL synthesis, 4-6, 10-11, 15, 41, 69

S

schematic capture, 2
scheduled design, 67, 74
scheduling, 17-18, 30, 120-121, 123
- algorithms, 30
- assumptions, 73-75, 181
- modes, see I/O scheduling modes
- of FIR filter, 268-269
- of JPEG compression algorithm, 256-257
- of Viterbi decoder, 281-282

sequential components, see pipelined components and memories
signed type, see data types
simulation, cycle-accurate, 39-40
state machine, 5, 18, 39, 79-80
- circular, 80
- Mealy, 82

std_logic / std_logic_vector type, see data types
std_logic_1164 package, 43
std_logic_arith package, 45
subprograms, 173, 178
  (see also functions and procedures)
superstate, 118
superstate-fixed I/O mode, 118-119, 121
- advantages and disadvantages of, 128
- coding style, 235-238
- handshaking, 192-194
- scheduling conditional branches in, 124
- scheduling loops in, 127

- testing designs scheduled in, 121

synchronous memory, see memory
synchronous reset, see reset
Synopsys®, 44
synthesized design, 61
system design, 9
system-on-chip (SOC) designs, 9

T

technology-independent descriptions, 67-68
technology-independent design, 3
test bench, 58-60, 122, 133, 197
- clock generation, 204
- for FIR filter, 267
- for JPEG compression, 252
- for Viterbi decoding, 279-290
- input and output processes, 205
- messages, 210
- reset generation, 204
- structure of, 202

text I/O, 210
throughput, 143-144, 147
timing diagrams, 70-73
type, see data type
type definitions, 178

U

unconstrained designs, 30
unsigned type, see data types

Index

V

variables, 74, 106

Verilog®, 3, 5

VHDL, 3, 5

Viterbi decoder case study, 274-276

W

WAIT statements, 106, 110, 111, 114, 124
- multiple, 70

WHILE loop, 89-90, 115
   (see also loop)
- translation of, 89

wrapper model, 60-63